D0400151

Climate Myths

Northbrae
Books

Also by John J. Berger

Nuclear Power—The Unviable Option: A Critical Look at Our Energy Alternatives

Restoring the Earth: How Americans Are Working to Renew Our Damaged Environment

Environmental Restoration Science and Strategies for Restoring the Earth (ed.)

The Ecological Restoration Directory of the San Francisco Bay Area (ed.)

Charging Ahead: The Business of Renewable Energy and What It Means for America

Understanding Forests

Beating the Heat: How and Why We Must Combat Global Warming

Forests Forever: Their Ecology, Restoration and Protection

Climate Peril: The Intelligent Reader's Guide to Understanding the Climate Crisis (forthcoming, 2013)

Climate Solutions—Turning Global Peril Into Jobs, Profits, and Prosperity (forthcoming, 2013)

Climate Myths

The Campaign Against Climate Science

John J. Berger

CLIMATE MYTHS: THE CAMPAIGN AGAINST CLIMATE SCIENCE

First published in 2013 in the United States by Northbrae Books.

Northbrae Books, Publishers
941 The Alameda
Berkeley, California 94707
Northbraebooks@gmail.com

ISBN: 978-0-98590-920-8 (paperback)
ISBN: 978-0-98590-921-8 (E-book edition)

Library of Congress Control Number: 2012913562

Library of Congress Cataloging-in-Publication Data

Berger, John J.
Climate myths: the campaign against climate science/John J. Berger
p. cm
Includes Appendix, Bibliographic References, and Index
ISBN: 978-0-98590-920-8 (paperback)
1. Climate science, denial of—United States. 2. Climate change, skepticism of—United States. 3. Climate change and fossil fuel industry—United States. 4. Climate change lobbying—United States. 5. Climate change disinformation—United States. 6. Climate science, myths about—United States. I. Title

2012913562

Cover design by Chii Maene
Author photo by Phil Saltonstall

Copies printed by Thomson-Shore of Dexter, MI 48130 contain post consumer recycled content with FSC-certified mixed sources. Paper consumption has been further offset through donations to support tree planting by Sempervirens Fund of Los Altos, California, an organization dedicated to restoration and protection of forest and park land, with additional trees planted by the author.

Manufactured in the United States of America
First edition January, 2013
1 0 9 8 7 6 5 4 3 2 1

To Dr. James Hansen and the other great and courageous scientists who have unraveled the mysteries of global climate change and had the courage to call attention to their findings.

“For the great enemy of truth is very often not the lie—deliberate, contrived, and dishonest, but the myth—persistent, persuasive, and unrealistic. Too often we hold fast to the clichés of our forebears. We subject all facts to a prefabricated set of interpretations. We enjoy the comfort of opinion without the discomfort of thought.”

– President John F. Kennedy
Commencement Address, Yale University
New Haven, Connecticut, June 11, 1962.

“ . . . The claim that global warming is caused by man-made emissions is simply untrue and not based on sound science.”

— Senate Floor Statement (July 28, 2003) by U.S. Senator James M. Inhofe (R-Oklahoma), then-Chairman, Senate Committee on Environment and Public Works, and author in 2012 of *The Greatest Hoax: How the Global Warming Conspiracy Threatens Your Future.**

“There are powerful voices of unreason, but every year, the science becomes stronger and the data are telling an ever more consistent story.”

— Dr. Benjamin D. Santer, research climatologist, the Lawrence Livermore National Laboratory’s Program for Climate Model Diagnosis and Intercomparison.

“If you look at what the tobacco industry was doing decades ago. . . . Now, we’re seeing the same thing with the fossil fuel industry trying to discredit scientists like myself linking the use of their product, fossil fuels, to the health of the planet.”

— Professor Michael Mann, Pennsylvania State University climate scientist and author, *The Hockey Stick and the Climate Wars: Dispatches From the Front Lines.*†

* Senator James M. Inhofe has called for an investigation of 17 climate scientists, including Professor Michael Mann.

† Interview with Dr. Mann, April 14, 2012 by ABC News Correspondent Bill Blakemore abcnews.go.com/blogs/technology/2012/07/new-mccarthyism-described-by-climate-scientist-michael-mann/ in *Nature’s Edge Notebook #31*, July 8, 2012.

Contents

Foreword

Scientists now clearly understand that humans are producing discernible climate change. There is no doubt. The planet is warming, and there are multiple signs, not just rising global mean surface temperatures, but also melting glaciers and Arctic sea ice, melting ice sheets (in Greenland and Antarctica), and rising sea levels, both from adding more meltwater from land and from the expansion of seawater itself due to rising temperatures.

These changes matter a great deal. Some of the changes with the biggest impacts on society and the environment are those related to increases in the intensity of heavy rains, droughts, and floods, and those associated with heat waves and wildfires. Changes in storms, especially hurricanes, are also in the cards.

The evidence for all these changes is widespread and incontrovertible, even if details still need to be ironed out. So why are there misinformation campaigns? Cutting back on coal and oil use would make these resources last longer, along with the fossil fuel industry that depends on them, and putting a price on carbon would surely get passed on to consumers. Also, it is not as if demand for power is diminishing—quite the contrary. So to me, the misinformation campaigns are not only dishonest and wrongheaded, but do not even serve their own industry well.

This book focuses on the myths perpetrated by so-called skeptics or deniers of climate change and the organizations and people who are responsible. In science, being skeptical comes with the territory, but to deny basic facts makes no sense at all. Hence, this book performs a service by providing a handy foil to the often silly arguments brought forth by climate change deniers. It also names names.

— Kevin E. Trenberth
September 27, 2012

Dr. Kevin E. Trenberth is a distinguished senior scientist in the Climate Analysis Section at the National Center for Atmospheric Research. For more information see page 105 or visit www.cgd.ucar.edu/cas/trenbert.html.

Preface

Climatologists, ecologists, and indeed all scientists who conduct research on the environment's capacity to maintain the quality of life for humanity are increasingly and sadly aware of three intertwined threats to human welfare. One is the harm being caused to air, water, soil, and life on Earth by such human actions as burning fossil fuels and abusing our land. The second is the ever-increasing number of human mouths to feed and thirsts to quench. But the third is the most dangerous, for it not only promotes the first two threats but also extends its reach to every facet of humanity's effort to govern itself rationally. It is the growing rejection of the premises, methods, and findings of science. John Berger's book, while concerned with climate science, serves a broader purpose by providing an antidote to a poisonous corruption of rationality that afflicts our society today.

Millions of Americans have doubts about whether global warming science is good science and a substantial proportion of our population has doubts about evolution. Now, denial of the findings of science can be a relatively harmless pastime provided it is done in the privacy of one's home. Indeed, the Constitution guarantees all of us the right to spout the most preposterous nonsense, so there will always be those who distort the findings of science. But it is a different matter altogether when creationists threaten to alter public school curricula, or when anti-science emanates from our legislative bodies and threatens to tear the legal fabric that protects our environment and society. In such instances, the citizens of a democracy *must* speak out in defense of rationality. *Climate Myths* does so to help ordinary citizens understand the contrived myths and manufactured controversy that opponents of climate science have propounded.

Some of anti-science's popularity derives from a general failure of U.S. science education. The problem is not a failure to provide scientific information to the public (we do plenty of that, and we do it well) but rather to inform the public about how science works. Thus deniers of global warming science say, "Global warming has not been proven—it's only a theory." Similarly, creationists delight in stating that, "Evolutionary theory has not been proven." This despite the fact that science can never prove anything—only mathematicians do proofs.

By contrast, in science we collect evidence and probe the natural world for testable ideas called hypotheses (or theories) that have predictive value. When the world is thus explained using these scientific processes, the results are not theories in a pejorative sense, but they represent the best provisional answers that science and society can presently provide. To disregard these findings in favor of the climate myths debunked in this book, for example, just because a more refined scientific theory may come along in the future, is the height of folly.

It is one thing for politicians and commentators who dislike the message of global warming to restrict their anger to the messengers—scientists. This is reprehensible, but is not the worst of the problem. Increasingly, those who hate the message are attacking not just the messengers, but the very language and rules of science (see Paul and Anne Ehrlich's *Betrayal of Science and Reason*,[a] or Naomi Oreske and Erik M. Conway's *Merchants of Doubt*[b]).

Science is built upon an ethical and logical framework comprised of peer review, controlled experiments, and the acceptance of, not the deliberate twisting of, a common terminology. Those who loathe the findings of global warming science or evolution have gotten away with undermining this framework because the general public is not familiar with the

---

[a] Paul R. Ehrlich and Anne H. Ehrlich, *The Betrayal of Science and Reason: How Anti-Intellectual Rhetoric Threatens Our Future* (Washington, D.C. and Covelo, CA: Island Press, 1998).

[b] Naomi Oreske and Erik M. Conway, *Merchants of Doubt: How a Handful of Scientists Obscured the Truth on Issues From Tobacco Smoke to Global Warming* (New York: Bloomsbury Press, 2010).

process of science. The public considers the word "theory" to be synonymous with idle speculation, even though, when scientists refer to the theory of relativity or of evolution, they are referring to our highest level of accumulated knowledge. Purveyors of anti-science prey on this confusion. *Climate Myths* helps dispel it by laying out the clear, irrefutable facts that science has provided about climate change.

To understand the damage, consider the alternative. Consider a citizenry that understands the nature of scientific evidence, that recognizes when it is conned by commentators, and that has the numeracy skills to know when it is being bamboozled by practitioners of mathematical malpractice. Such a citizenry is *not* going to allow governing bodies to be guided by anti-science. It would not allow their government to continue to override the scientific findings of the nation's most distinguished scientists—as many of our influential politicians have been doing over global warming, arsenic in drinking water, forest protection, and many other issues.

When political leaders rely on quackery to bolster their case to overturn the rules and regulations that protect citizens from polluting and looting our natural heritage, they are really not merely overturning a legislative legacy of several decades. They are doing something even more damaging—overturning a 500-year legacy of civilization that has steadily replaced faith in irrationality with understanding of, and acceptance of, the process of science. The long-term consequences of such actions will reverberate through the coming decades in ways that will degrade far more than the health of our environment and our material well-being; they will erode the very foundations of our democratic society.

Two hundred years ago, Thomas Jefferson understood this fragility of democracy well. He knew that an illiterate citizenry would be unable to sustain the institution of democracy and for that reason, he greatly expanded access to public education. With the passage of 200 years and the ballooning of technological capacity, it is equally important now that citizens possess a basic acceptance of the methods and findings of science.

Many interest groups are pushing in the opposite direction, toward an "Age of Endarkenment." At the forefront are those who simply cannot accept the authority of science and who loathe having government take the steps needed to heal the environment, even when those same steps are probably also needed to heal our economy and sustain our future. The reason appears to be flat-out hatred of the concept of government. This is individualism turned cancerous.

The problem is not religion. I returned recently from two months of sabbatical research and travel in three countries: Turkey, a largely Muslim country, Bhutan, a Buddhist nation, and Andalucia in Spain, a largely Catholic nation. The overall understanding of global warming amongst the admittedly small cross-section of people I spoke with was refreshing and consistent. On the world stage, where international negotiations over climate treaties take place, and at the local level, where societal support for clean technology matters, these nations look much more like leaders than do we. *Climate Myths* dispels the myth that adhering to international climate treaties is a forfeiture of national sovereignty, and shows why the U.S. ought to take a leadership role in promoting an effective international climate treaty.

John Berger's useful book explains the science of global warming and debunks the myths created by those who deny the relevant science. Another useful, indeed essential, task is to better communicate to the public the ethical foundation of science. Some call this foundation "the scientific method." But the stodgy manner in which "method" is explained in high school textbooks tends to be off-putting. The scientific method is based on ethical premises, but we don't seem very good at talking about them. Perhaps we need to spend more time describing how the sheer joy of scientific discovery, and the beauty it reveals about the natural world, stem less from individual brilliance than from the collective acceptance by a community of scholars of a set of rules, a scientific ethic, that constitutes not merely a "method," but more importantly a covenant that has been empirically reinforced over 500 years of human history, and is now under severe attack by opponents of science and by climate science deniers.

— John Harte, Berkeley, CA
October, 2012

***Dr. John Harte** holds a joint professorship in the Energy and Resources Group and the Ecosystem Sciences Division of the College of Natural Resources at the University of California, Berkeley. He has authored over 190 scientific publications, including eight books, on topics including biodiversity, climate change, biogeochemisty, and energy and water resources. For more information, see page 104 or visit: erg.berkeley.edu/people/faculty/John_Harte/harte.shtml.*

Introduction

There are still people who deny that our climate is changing. The climate, though, just won't cooperate.

The summer of 2012 was the hottest ever for the continental United States. Half of the nation's corn crop was wiped out, along with 60 percent of our pasture land, by the worst drought in half a century. Wildfires scorched nearly nine million acres of forests and fields. Arctic ice melted to its lowest level on record. And carbon levels are rising so rapidly in our oceans that the world's coral reefs, and all they support, have become imperiled, perhaps beyond their ability to ever recover.

We're thus now actually seeing what climate scientists have been warning us of for decades. This is what climate change looks like. This is a calling card from the future we face unless we act now to turn it around. This is a taste of the grim reckoning that lies ahead if we remain on the road to fossil fuel ruin.

Most Americans have gotten the message. 74 percent of the country believes the federal government should regulate greenhouse gases to reduce global warming, according to an August 2012 *Washington Post*/Kaiser Family Foundation poll that surveyed 3,120 Americans nationwide.

And yet, no matter how hot and bothered the planet and the public seem to be, our political climate remains unmoved. Comprehensive energy and climate legislation failed two years ago in the Democratic-controlled Senate. And the Republican-controlled House of Representatives has voted repeatedly since early 2011 to block efforts to roll back the carbon pollution that is warming our Earth and threatening us all.

There's no basis in science for the political paralysis. Indeed, among those who best understand the science of climate, global warming and its causes are not in doubt.

> "Climate change is occurring, is very likely caused primarily by the emission of greenhouse gases from human activities, and poses significant risks for a range of human and natural systems. Emissions continue to increase, which will result in further change and greater risks. Responding to those risks is a crucial challenge facing the United States and the world today and for many decades to come."

So sayeth the National Academy of Sciences' 2011 report, *America's Climate Choices.* It is important to understand that the National Academy of Sciences was created by the U.S. Congress "to investigate, examine, experiment, and report on any subject of science or art" at the request of our government. It was chartered during the American Civil War, a moment in history when our leaders were serious about needing to understand the bedrock truth about our physical world, as best as can possibly be known. This is our national brain trust.

The climate report pulled together hundreds of peer-reviewed analyses and exhaustive studies. It was developed under the supervision of twenty-two senior scholars in their field from places like Princeton, Georgia Tech, the Massachusetts Institute of Technology, Stanford, and Duke. There is no more qualified group of experts to pass judgment on known science. There is no more authoritative approach for discerning consensus. And there is no more unassailable assemblage of the intellectual and probative critical mass required to reach a definitive and actionable conclusion.

The action we need is no mystery. We burn, in this country, 800 million gallons of oil every single day. That's enough to fill the Empire State Building three times. We get 35 percent of our electricity from coal, the single greatest source of the industrial carbon pollution that is wreaking havoc on our climate. We need to invest in energy efficiency, so that we can do more with less. We need to develop wind, solar, and other sources of clean, safe, renewable power. We need to cut the carbon emissions from our factories, trucks, and cars. And we need to reduce our reliance on fossil fuels, not overnight, but over time.

By doing so, we can strengthen our economy, put our people back to work building the next generation of energy efficient cars, homes and workplaces, and position American companies for success in the growing global market for the environmental and energy solutions of tomorrow. We'll make our country more secure, by reducing our reliance on foreign oil. And we'll create a healthier future for our children by at last addressing the epic scourge of climate change. What other policy strategy offers so much benefit to the country and so much potential for our future?

And yet, despite the imperative to act and the opportunity that taking action would present, we find ourselves at a point where no functional democracy should ever arrive. Our congressional leaders cannot be swayed by empirical science of the highest order. They will not be moved by the will of the people, nor by the public good. They will not act on the plain and immutable truth that is, literally, seared across the heartland.

Why, in heaven's name, not?

John Berger has answered that vexing question in this small but important book. His conclusion will outrage every American with a political conscience, a dedication to our future, or a sense of simple decency.

Our political process has been hijacked by the fossil fuel industry for no higher calling than profits, measured in the scores of billions of dollars each year. The reason our politicians have failed to act—and in so failing, have failed the country—is that the oil, gas and coal companies, along with the electric utilities, cement factories, and industrial incinerator operators that burn fossil fuels, do not want them to act.

These top-line polluters, Berger explains, have too much at stake to risk any action that might result in a reduction of our dependence on fossil fuels. And so they spend hundreds of millions of dollars, every year, lobbying our elected officials in Washington and pumping up the campaign coffers of anyone who can be counted on to say no to action on climate change. Climate change denial, in fact, has become a kind of litmus test for roughly half the members of Congress. Stand up for our future, speak out against carbon pollution, tell the truth about what's happening to our climate and putting our future at risk, and a politician flunks the litmus test. Yes, Berger makes clear, it really is that simple.

But there's more to the dark bargain than back room arm-twisting and influence peddling. The fossil fuel industry provides these politicians cover by investing heavily in public deception. They pay second-rate hacks to trump up junk science. They hire Madison Avenue public relations flacks to put together the same kind of slick advertising campaigns the tobacco industry once used to fool customers into believing they weren't killing themselves by smoking. They launch chilling and sometimes debilitating attacks on scientists working to advance the march of knowledge.

The industry's goal isn't to make the public believe climate isn't changing. It's the far simpler task of merely sowing enough doubt to give the fossil fuel allies on Capitol Hill a politically plausible excuse for inaction. Why lose out on fossil fuel funding, after all, and risk being targeted by industry-funded Super PAC attacks, when you can simply shrug your shoulders and tell constituents that more study on the matter is needed?

The strategy has been as successful as it is disgraceful, Berger argues in clear and cogent prose supported by extensive research and documented facts.

Hardly an ideological flame-thrower, Berger is a noted scholar, a graduate of Stanford University with a master's degree in energy and natural resources from the University of California, Berkeley, and a Ph.D. in ecology from the University of California at Davis. He has authored or edited 11 books on climate, energy and natural resources, including an invaluable forthcoming 2013 resource, *Climate Peril: The Intelligent Reader's Guide to the Climate Crisis.*

What Berger has done in the current tightly written volume is to rip the veil of respectability off the face of climate denial. This isn't about science. It isn't about fact finding. It isn't about what's best for our children's future. It's the opposite of all that.

Climate denial is an industrial conspiracy to defeat science, refute fact, and make a mockery of any ambition to leave behind a better world for those who follow us. If the effort to block action on climate change continues to succeed, the result will be large and growing swaths of Earth that will be uninhabitable for the next generation, and for all generations to follow.

Berger has laid out the problem before us. The fix is within our grasp.

The fossil fuel industry, powerful as it is, cannot match the concerted voice of the American people. The industry has the right, in this country, and certainly the means, to spend hundreds of millions of dollars every year pressing its case. And we have an obligation to stand up and say what's best for the rest of us. We have the power to bring about change. The time to begin is now.

— John H. Adams,
Beaverkill, N.Y.
October, 2012

John Adams is the founding director of the Natural Resources Defense Council, a national environmental advocacy organization with more than 1.3 million members and supporters nationwide. He received the Presidential Medal of Freedom in 2010 in recognition of four decades of service to the nation as a voice for environmental protection. *See page 103 for more information.*

CHAPTER 1

CLIMATE MYTHS: THE CAMPAIGN AGAINST CLIMATE SCIENCE

The oil and coal industries—and many of their big industrial customers—have for decades tried to dismiss concerns about climate change, discredit climate science, and thwart climate legislation and policy. Working through pseudoscientific think tanks, institutes, and assorted right-wing proxy organizations, they manufacture myths about climate change, paralyze the climate policy process, criticize reputable scientists, and pretend that policies to protect the climate pose a dangerous threat to democracy and prosperity. But as you will see, such myths do not stand up to scrutiny. The pages that follow spotlight the companies, organizations, and individuals conducting the highly successful multimillion dollar disinformation campaign that has set back global climate protection by decades that the world can ill afford to waste. You will learn the tactics and strategies that have been used to effectively mislead tens of millions of people around the world.

* * *

CLIMATE DISINFORMATION

For decades, the oil and coal industries and some of their largest industrial customers have conducted a sophisticated and wildly successful multimillion dollar campaign based in the U.S. to convince the American public that climate change is not a serious threat. The impetus for the campaign has been to protect industry profits by blocking any action designed to reduce the emission of carbon dioxide and other global heating gases produced in burning fossil fuels.

Policies such as carbon taxes and carbon caps are intended to limit the release of carbon dioxide by restraining demand for fossil fuels. Fossil fuel companies, however, have correctly concluded that crimping fuel consumption would reduce revenue and would also erode the multitrillion dollar value of their oil, coal, and gas reserves.[a]

Fossil fuel industry leaders have long known that as policies to address the dangers of fossil fuel burning and climate change were progressively made into law and policy, they would ultimately affect profits. Anticipating these threats to their income and wealth, large fossil fuel energy companies—and those who have made common cause with them—decades ago mounted a well-funded campaign to discredit climate science. Its architects recognized that, if successful, the campaign would provide the rationale for their political and legislative efforts to obstruct public policy efforts aimed at climate protection. The beginning and expansion of this campaign are the subjects of this book.

While the campaign has served and continues to serve a political and economic purpose for the industries behind it, it also serves the psychological need of reconciling industry's economic interests with their version of climate science, climate economics, and the economics of climate protection. Thus those in the climate science denial camp believe themselves "on the side of the angels."

In the political arena, the energy company campaign not only succeeded in confusing facts about climate change but also managed to undermine U.S. participation in the 1997 Kyoto Protocol, a precedent-setting international climate protection treaty discussed frequently throughout this book (see index for

[a] A wholesale write-down of these reserves will occur when financiers and other investors decide that a portion of these reserves cannot be developed. These reserves contain several times as much carbon dioxide as the atmosphere holds and far more than the 565 gigatons that scientists believe it can contain before global temperatures spike more than 2° C. The complete combustion of these reserves would thus destroy the Earth's climate. Therefore, much of these assets inevitably will have to be entirely written off for the foreseeable future.

complete listing). The fossil fuel industry achieved this political triumph by providing the arguments that were used in turning Congressional sentiment against the Protocol.

The industrial opposition to climate science and climate-safe energy policies has grown more sophisticated and varied over the past decade. The campaign operates through dozens of industry-funded institutes, policy centers, councils, research foundations, and societies that speak for industry on climate and energy.

The climate "skeptics," as they like to be called, include anti-government and anti-regulation conservatives and libertarians who oppose government action on ideological grounds. Their strategy has often been to hide ideologically based misrepresentations of climate science beneath a mantle of science.

A review of scientific publications on climate, however, reveals that whereas many thousands of high-quality scientific papers validated by peer review have been published documenting all phases of global warming, only a trivial number of dissenters who dispute the evidence have published in similar journals.

Moreover, by contrast, the results of climate studies confirming global warming and humanity's role in it can be found in the most prestigious scientific journals. Almost without exception, the deniers' reports appear in publications that are not peer reviewed, since their objections to climate science have been repeatedly refuted; thus they are of little interest to responsible, well-respected scientific publications. Finally, the national academies of science of most nations of the world have passed resolutions affirming that we are warming the planet.[1]

Although climate change is a scientific issue, it has been adopted as a Republican "litmus test" issue by certain Republican Party spokesmen and thus public opinion surveys show that more Republicans than Democrats characterize themselves as "climate skeptics." These individuals today appear less focused on disparaging climate science than in the past, when climate science was less settled. Nowadays they seem to have shifted tactics to focus more attention on defeating the environmental and energy policies implied by climate change concerns.

In the initial stages of the climate debate, industry proxy organizations often flatly contradicted climate science and claimed, variously, that the Earth was cooling or at least wasn't warming, or that if the Earth was warming, the warming wasn't due to human activity, or that if the Earth were to warm, it would be mild and beneficial.

Many of these discredited claims have been abandoned by all but diehard opponents of climate science as the global scientific consensus on climate change has strengthened and as the evidence for global warming has become overwhelming. Some deniers still persist in presenting discredited arguments, however.

For example, industrial critics of decisive action on climate change (such as the National Association of Manufacturers) made a case in Congress and with the public in 2009 that effective measures to reduce carbon emissions would bring economic disaster in the form of high taxes, lost jobs, lower productivity, and reduced competitiveness for America in world markets.[2]

Since their arguments weren't gaining traction in the world of science, industry-funded think tanks then spent millions of dollars making their case against climate science to more gullible media, government officials, opinion leaders, students, and the general public. Climate skeptics and their allies have thus become a major presence on the Internet, over radio, and on TV airwaves, as well as through industry-sponsored books, magazines, articles, reports, and press releases.

An unsuspecting person who uses an Internet search engine and enters terms commonly associated with climate change will be hard pressed to discern the truth amid the plethora of misleading information many of these organizations provide. Since some of the most effective arguments consist of deceptive statements wrapped in layers of truth, it can be very challenging for students and others without advanced scientific training or sophisticated rhetorical and analytical skills to sift truth from falsity without investing lots of time.

To put climate skeptics' current claims in perspective, it is useful to review the initial 20th century phase of the energy industry's assault on climate science and its related effort to block Federal laws and policies to combat global warming. I will then examine some of the main myths they manufactured about climate change and explain their fallacies.

At this point in the narrative, readers who are unfamiliar with climate science and why the release of greenhouse gases disturbs the climate can skip directly to page 35 of this book to read the point-by-point discussion of climate myths. Readers can also soon consult a companion volume of mine, *Climate Peril: The Intelligent Reader's Guide to the Climate Crisis* (Northbrae Books, 2013).

SABOTAGE OF THE KYOTO PROTOCOL

As noted earlier, the industries trying to defuse concern over global warming were successful in mobilizing opposition to the 1997 Kyoto Protocol. The Kyoto agreement was modest in its goals and was meant as an inoffensive and attainable first international step toward controlling global carbon emissions. Thus the Kyoto Protocol called on the industrialized nations to reduce their greenhouse gas emissions by only five percent by 2012, instead of the nearly 90 percent or more needed to stabilize the atmosphere's carbon dioxide (CO_2) concentration (see pages 52-53).

The U.S. signed the agreement, although it had already been weakened during negotiations, in deference to U.S. industry opposition. Industry was then successful in winning Senate support (in a 95 to 0 vote) for a resolution declaring that the

U.S. should not be obligated to cut its carbon emissions unless developing nations were also obligated to do so.

Because the Kyoto Protocol had no provisions requiring mandatory emission reductions by developing countries, the Clinton Administration knew the Kyoto Protocol would be defeated in the Senate and thus did not even submit it to the Senate for ratification. President George W. Bush, President Clinton's successor, withdrew the U.S. from the Kyoto agreement, even though, as a presidential candidate, he had said that he would cap carbon emissions if elected.[3]

Of course, in a Catch-22, the developing nations of the world that have historically contributed relatively little atmospheric carbon dioxide, and who are far less affluent that the U.S., clearly would not want to reduce their carbon output before the U.S. agreed to reduce its own.

The forces behind the anti-Kyoto campaign included the coal, oil, auto, electric, metal, chemical, paper, cement, and railroad industries. Their efforts were joined by various anti-environmental groups. Members of this coalition seemed to take it on faith that the unrestrained use of fossil fuels was a good thing and that continual economic growth was possible and desirable, without drawbacks. I call this coalition "the fossil fuel industry" because of their common interests in continuing very heavy reliance on fossil fuel. (The natural gas industry does not march in lockstep with the oil and gas lobbyists on climate issues, for reasons to be explained shortly.)

Why have so many multibillion dollar industries opposed a consensus about climate change forged by the world's leading climate scientists? The executives of these companies live on this planet, too. They have children and grandchildren. But since they profit directly from either the production or use of carbon-based fuels, or both, they clearly do not want their costs increased or the value of their assets decreased, nor do they want to see their hundreds of billions of dollars in sales revenue reduced.

In particular, they regard the possibility of significant new energy taxes, which would make fossil fuels more expensive, a serious threat to their prosperity. With trillions invested in fossil fuel energy infrastructure, it is no surprise that these industries are not exactly keen on renouncing the carbon economy.

Despite the obvious public health benefits that fossil fuel industry leaders and their progeny would receive if we protected the climate,[b] the fossil fuel industry takes a dim view of stronger clean air standards that impose pollution control costs on fossil fuel polluters. Even if those costs are passed on to consumers, higher costs restrain demand and siphon away revenue. Technologies that improve fuel efficiency also depress fuel demand. So the fossil fuel industry has a powerful vested interest in opposing these antidotes to climate change.

NIXING A CARBON TAX

The fossil fuel industry and its allies have a long history of working effectively together to oppose clean air and water regulations and fuel efficiency standards. During much of the 1990s, these industries spent millions of dollars hiring public relations firms, sponsoring industry-funded think tanks, and creating corporate coalitions with misleading names that sounded like grassroots organizations but did industry's bidding.[c]

Using these surrogates, the oil and coal industries and their allies sent their message out through press briefings, web sites, news releases, newspaper ads, conferences, email and petition drives, and direct contacts with editors, columnists, television correspondents, government officials, and business leaders. In 1993, they halted the Clinton Administration's efforts to pass a modest tax on the energy content of fuels, which would have helped control carbon emissions. This victory set back the nation's efforts to control its emissions by at least 20 years.

[b] See *Solving the Climate Crisis*, cited earlier.

[c] Some of the public relations experts hired had previously labored for the tobacco industry to downplay the risks of cigarette smoking.

Later, these industries lobbied Congress to oppose domestic renewable energy and energy efficiency programs, which they called back-door implementation of the Kyoto Protocol.

ORIGINS OF THE CLIMATE "DEBATE"

Since they were not able to disprove any important aspects of climate science, the main tactics of the fossil fuel industry's climate campaign have been to sow doubt about climate science and to sow fear that efforts to reduce global warming will inflict grave economic pain. They soon found that it was easy to create doubt by raising sensationalistic questions, even if the questions had no scientific merit, so long as they required time and expert knowledge to rebut. That would be enough to make the public uneasy, unable to dispel the doubts raised.

The industries recognized many years ago that even if they were unable to refute climate science, they could achieve their objectives of delaying and obstructing government action for years by raising enough doubts to stir up opposition and paralyze decision makers.

These doubts were relentlessly spread to Congress year after year through briefings and other contact with Congressional staffers, supported with reams of biased literature and by misleading testimony from scientists recruited to the industry's public relations campaign. Few Congressmen had sufficient scientific training to see through this ruse, perpetrated with the help of scientists who received compensation from industry.

This strategy was very effective in preventing Congressional action on climate change, since it convinced many Senators and Representatives that climate science was fraudulent or at least controversial and hence too unreliable to base policy and law upon.[4]

Once industry created enough doubt in the public's mind and sowed enough fears, it did not need to prevail on the merits of the scientific debate. They won their short-term victory and thus delayed meaningful action to reduce carbon emissions and protect the climate.

Fortunately for industry, it is far easier to make unfounded charges and raise misleading questions than it is to authoritatively refute them. Reasonably intelligent public relations operatives and clever lobbyists can endlessly raise invalid but plausible-sounding issues that the public has great difficulty in distinguishing from genuine scientific controversy.

In the next section, I'll summarize the coal and oil industries' early disinformation campaign against climate protection policy. Then I'll review the stands of the right-wing and libertarian think tanks that profess nonpartisan educational aims yet promote industry views. Finally, I'll examine some of the industry's most basic climate change myths.

INDUSTRY'S SPIN ON CLIMATE CHANGE

The fossil fuel industry often works through proxy organizations and individual climate skeptics, who generally have no credibility on climate issues but who are good at shaking public confidence in the conclusions of climate science to paralyze the policy making process.

The Greening Earth Society was one such organization. You might think from its name that it was an environmentally oriented group. But no, The Greening Earth Society was a creation of the Western Fuels Association, a $400 million coal producer co-op.

From its website (www.greeningearthsociety.org), this benevolent-sounding "green organization" served as a gateway to coal, oil, and mining industry-funded think tanks and institutes as well as to publications rife with misinformation. Some of the materials circulated by the "Just Say 'No' to Climate Change" folks even targeted elementary school children through their teachers.

When I last searched for www.greeningearthsociety.org in 2008, the website was no longer operational. However, at the Western Fuel Association's website, I found a link to the Center for the Study of Carbon Dioxide and Global Change, whose website denies a connection between the Earth's recent warming and increased atmospheric carbon dioxide.

C.D. Idso and K.E. Idso's 1998 treatise, "Carbon Dioxide and Global Warming: Where We Stand on the Issue" was prominent on the Center's website, www.co2science.org, in June 2008. "Atmospheric CO_2 enrichment brings growth and prosperity to man and nature alike," they claimed; translation: global warming is good for nature and humanity. Co-author Craig D. Idso, is the Center's founder and former president and he's the former Director of Environmental Science at Peabody Energy Company. Peabody is the world's largest private coal company, fueling 10 percent of all U.S. electricity generation.[5]

Climate skeptics have played a critical role in the coal and oil industries' efforts to foster doubts about climate science and fears of an economic meltdown. Although the skeptics present themselves to the public as independent scientists or respected climate experts, most of the best known of these "objective thinkers" have taken significant amounts of energy industry money for themselves or their organizations, and they espouse scientifically dubious positions.

Prominent examples include Dr. S. Fred Singer, funded at times by Exxon, Shell, Unocal, ARCO, and Sun Oil; Dr. Pat Michaels, recipient of at least $165,000 from coal and other energy interests; Dr. Richard Lindzen of MIT, who has received money from the Western Fuels Association; and climatologist Dr. Robert Balling of Arizona State University, whose work received over $300,000 from coal and oil interests.[6]

Individuals like these, supporting views far outside mainstream climate science, have paraded before the media, their presence falsely suggesting a pervasive disagreement among climate scientists and obscuring their wide consensus. At times, climate skeptics have recycled discredited scientific opinion on the assumption that the public would be unable to sort out the truth.

In doing so, they enjoyed a great advantage. Unwary or irresponsible members of the press have often given these erratic views equal time with those of responsible, reputable climate scientists, creating the false impression that the basic ideas of climate science are widely disputed. Uninformed readers and listeners might be inclined to regard both sides of the make-believe controversy as equally credible, and "split the difference," since one side said there was a serious problem and the other side claimed there was none.

For an example of just how irresponsible a newspaper can be in publishing nonsense about climate change, see, "Science Has Spoken: Global Warming is a Myth," which appeared in *The Wall Street Journal* on December 4, 1997. Its authors, chemist Arthur Robinson and his son Zachary, ran the tiny Oregon Institute of Science and Medicine outside Cave Junction, Oregon, from which they marketed nuclear bomb shelters and home-schooling advice.

Relying on the mistaken claim that changes in solar activity explain the Earth's increase in temperatures since the Little Ice Age, the article concludes, "There is not a shred of persuasive evidence that humans have been responsible for increasing global temperatures." The article then advises readers not to worry "about human use of hydrocarbons warming the Earth."

"Carbon dioxide emissions have actually been a boon for the environment," the article states. "Our children will enjoy an Earth with twice as much plant and animal life as that with which we are now blessed. This is a wonderful and unexpected gift from the Industrial Revolution."[7]

Another attempt to cloak the climate disinformation campaign in the trappings of science was a "Global Warming Petition" supposedly signed by 17,000 U.S. scientists, but whose names were published without any identifying titles or affiliations. (The list included author John Grisham, several actors from the TV series M*A*S*H*, and a Spice Girl.) The petition was circulated by none other than Dr. Robinson's Oregon Institute of Science and Medicine.

With the petition came a bogus "eight-page abstract of the latest research on climate change," formatted to look like a published scientific article from the prestigious *Proceedings of the National Academy of Sciences,* with which it had no connection.

Filled with misinformation and put together by the Robinsons and two coauthors affiliated with the George C. Marshall Institute, the tract was accompanied by a letter of endorsement from the late Dr. Frederick Seitz, a former president of the National Academy of Sciences in the 1960s, who contended that "global warming is a myth." Dr. Seitz was a physicist, not a climatologist, and in the opinions of at least two very prominent scientists, "has no expertise in climate matters." He had been, however, "one of the last remaining scientists who insist that humans have not altered the stratospheric ozone layer, despite an overwhelming body of evidence to the contrary." Dr. Seitz's views illustrate that expertise and professional distinction, even in physics, does not insure good judgment in another area of science and policy.

Chapter 2

Think Tanks, Foundations, and Other Campaign Allies

A Chorus of Skeptics

Tactics used in the coal and oil industries' challenge to climate science can be traced back at least to the formation of the Information Council on the Environment (ICE) in 1991 by the National Coal Association, the Western Fuels Association, the Edison Electric Institute, and others.

Bracy Williams & Co., a Washington, D.C. public relations firm, was hired to run the organization with a $500,000 budget for advertising and public relations. The firm assembled a "scientific advisory panel" consisting of climate skeptics to further ICE's self-described mission: to "reposition global warming as theory (not fact)." But because internal memos were leaked, the ICE campaign was aborted.

The demise of one industry-funded front organization, however, did little to deter the fossil fuel industry from pursuing its long-term goal of stalemating action on climate change. They merely advanced their agenda through other channels.

Perhaps the most important of these was the Global Climate Coalition (GCC), set up in 1989 by the public relations firm of Burson-Marsteller for the fossil fuel industry and its industry allies. Meeting at the offices of the National Association of Manufacturers, this influential coalition was the apparent hub of the fossil fuel industry's climate campaign for years. From 1994 at least through 1998, the GCC spent upwards of $1 million a year to promote its climate views.

GCC members included the Aluminum Association, the American Forest and Paper Association, the American

Petroleum Institute, ARCO, Chevron, Cyprus AMAX Minerals, the Edison Electric Institute, Exxon, Mobil (Exxon and Mobil later combined to form ExxonMobil), Goodyear Tire and Rubber, the Southern Company, Texaco, railroads, and some 40 other companies. (By 2000, former members BP-Amoco, Shell, Chrysler, Dow Chemical, General Motors, and Ford had abandoned the coalition.)

Additional public relations services were bought from E. Bruce Harrison Co., which had been known as a leader of the pesticide industry's attack on author Rachel Carson and her book *Silent Spring.* (The book sounded an overdue alarm in the 1960s about the environmental harm caused by pesticides.)

According to an article by Chris Mooney in the May/June 2005 issue of *Mother Jones* magazine ("Some Like It Hot: As the World Burns"):

> "Drawing upon a cadre of skeptic scientists, during the early and mid-1990s, the GCC sought to emphasize the uncertainties of climate science and attack the mathematical models used to project future climate changes. The group and its proxies challenged the need for action on global warming, called the phenomenon natural rather than man-made, and even flatly denied it was happening . . . after the Kyoto Protocol emerged in 1997, the group focused its energies on making economic arguments rather than challenging science."[1]

Another important communications arm of the oil and coal industries was the Global Climate Information Project, founded in 1997 and guided by Shandwick Public Affairs, a Washington, D.C. public relations firm that at one time had a $13 million war chest for its climate change advertising campaign. The Global Climate Information Project promptly spent more than $3 million on ads designed to instill public fears that the Kyoto

Protocol would add a half-dollar tax to the price of gasoline while raising prices on everything else.

The large automobile manufacturers used similar tactics through their Coalition for Vehicle Choice (CVC), which paid for an expensive ad campaign to reinforce the idea that the Kyoto accord would hobble the U.S. economy.

Several other industry-funded groups have been active in disseminating allegations about climate science, climate change, or the costs of mitigating (moderating) its effects.

- **The American Enterprise Institute** is one of the nation's most powerful and prominent conservative think tanks, with conservative research fellows such as former Assistant Secretary of Defense Richard Perle (a proponent of the Iraq war) and Reagan Administration senior policy adviser Michael A. Ledeen (no longer with AEI). According to the organization's website, "AEI's purposes are to defend the principles and improve the institutions of American freedom and democratic capitalism—limited government, private enterprise, individual liberty and responsibility. . . ." Its Board of Trustees reads like a *Who's Who* of America's leading corporate CEOs and chairmen. The board was formerly chaired by (now-retired) ExxonMobil CEO Lee Raymond, and the Institute received $960,000 from ExxonMobil between 2000 and 2003.[2] Whereas in the past, some AEI fellows sought to cast doubt on climate science, AEI's climate papers stress the costs and difficulties of avoiding climate change while mostly avoiding direct challenges to mainstream climate science. AEI fellows attacked the Kyoto Protocol and continued to attack the value of cap-and-trade emissions controls[3] (see pages 21 and 31) while advocating that the nation adopt a mild carbon tax,[4] decades of research and development on carbon-free energy sources, and an emphasis on long-term solutions: ". . . [W]hat matters are global emissions levels 30, 50, and 75 years from now," wrote AEI fellow Samuel Thernstrom in AEI's magazine.[5]

As a wealthy multi-issue organization with a powerful corporate board and significant media access, AEI fits the classic model of the conservative policy institution in the National Committee for Responsive Philanthropy's report, "$1 Billion for Ideas: Conservative Think Tanks in the 1990s."[6] The institute disseminates its views through media coverage of its sponsored research and through its own conferences, books, magazines, and other publications. Its target audience is government officials and legislators, teachers and students, business executives, professionals, journalists, and the public.

- **The Cato Institute** is a libertarian think tank that has received support from the oil, gas, and chemical industries, the tobacco industry, and conservative foundations. It serves as a base for Dr. Patrick Michaels, a prominent climate skeptic.[7] Steven J. Milloy, founder and publisher of www.junkscience, which has scoffed at global warming, has also been listed as one of the Institute's adjunct fellows. He has been a source of climate skepticism for *The Washington Times*, Fox News, the Heartland Institute, and the Advancement of Sound Science Center, another ExxonMobil-funded group.

- **Consumer Alert** was funded by big oil and chemical corporations, among others, and operated a coalition of twenty-four nonprofits known as the National Consumer Coalition (NCC). A "Cooler Heads Coalition" within the NCC served as the group's voice on climate issues.[8] Both NCC and Consumer Alert reportedly ceased operations in 2005.

- **The Competitive Enterprise Institute**, which has supported the anti-environmental "Wise Use" movement, set in motion a series of anti-Kyoto conferences in 1997. "The Costs of Kyoto," the first of these conferences, provided a platform for attacks on climate science and for predictions that the Kyoto Protocol would be an economic disaster. The Institute is a major player in the effort to discredit mainstream climate science and received $1,380,000 from ExxonMobil from 2000 to 2003, according to, "Put a Tiger in Your Think Tank," a pithy exposé of the global warming disinformation campaign.[9]

  Posing the challenge of global warming as if it were an ideological issue, CEI President Fred L. Smith, Jr. characterized the global warming debate as "cultural warfare against economic liberty" in his foreword to Czech President Václav Klaus's anti-environmental *Blue Planet in Green Shackles:* "For that reason," Smith wrote, "pro-freedom voices are needed to reframe the debate to show how a free people can better address the challenges facing Western civilization."

  The premise of *Blue Planet in Green Shackles* is "that policies being proposed to address global warming are not justified by current science and are, in fact, a dangerous threat to freedom and prosperity around the world."[10] This appears to be an effort by the fossil fuel industry's defenders to politicize and emotionalize the climate issue, laying the groundwork for future attacks on people concerned about climate change. These tactics are not unlike those of the anti-communism crusades launched in the 1950s by the late and disgraced Senator Joseph McCarthy, an archetypical anti-communist witch hunter.

- **The Environmental Conservation Organization**, another bastion of the "Wise Use" movement, has supported conspiracy theories featuring the United Nations and environmentalists' designs to create a "one-world government." No sign of this organization was found in a 2008 online search.

- **The Foundation for Research on Economics and Environment**'s website, www.free-eco.org, informs readers that "the truly serious consequences of climate change will not appear for many decades," and that, "because the world will be much wealthier and more technologically adept in several decades, the best climate-change policy is to emphasize present economic growth, especially in the developing world."[11]

  FREE Chairman John A. Baden, Ph.D. implies that global warming is good in, "Join the Climate Change Crusade?"[12]: "Nearly all of humanity has benefited substantially during the past 10,000 years of warming and glacial melting. Europe flourished during the medieval warmth of 1000 to 1400, when the Vikings settled Greenland." He goes on to state, "If the projected warming of the next two centuries occurs, Russia, Canada, and parts of the U.S.A. are likely to prosper while other places decline or disappear underwater. It's naïve to expect those who expect huge gains from GW [global warming] to support costly initiatives to stop and reverse it." In this simplistic forecast of prosperity for countries in northern latitudes, Baden neglects to mention aboriginal peoples who will lose their livelihoods and lives, Arctic ecosystems and their wildlife, which will vanish, thawing permafrost that will release methane and exacerbate warming, seas that will rise, flooding of coastal areas, and damage to oceans, marine and coldwater fisheries, and to the Earth's shared natural resources, including its biodiversity. Reading this organization's website brings to mind the saying: "Economics without ecology equals nonsense." Baden does not seem to understand, that, as former senator and Wisconsin governor Gaylord Nelson once said, "The economy is a wholly owned subsidiary of the environment."[13] He seems oblivious to the fact that ecological damage on this scale, and wiping out vast numbers of species in a global extinction spasm, will on balance produce misery rather than prosperity. In June 2008, FREE's website contained an article by Baden titled, "The Skeptical Environmentalist,"[14] in which he mistakenly concluded: "We will have considerable time to adjust [to climate change]. Unless drastic as a new ice age, climate change is of little direct importance to people in the developed world. Wealth provides great resiliency. Aside from agriculture, *there is no significant economic activity much affected by climate, certainly not by the relatively minor changes scientists anticipate during the next century.*" [emphasis added].

- **The Frontiers of Freedom Institute** opposes environmental regulations, such as the Endangered Species Act, and some years ago co-sponsored "Countdown to Kyoto," an anti-Kyoto conference in Australia that showcased prominent climate skeptics. Frontiers of Freedom's website featured a *Wall Street Journal* opinion piece by U.S. Senator James Inhofe (R-Oklahoma), who has called global warming a hoax. In hyperbolic language, Inhofe charged on the Senate floor in June 2008 that the proposed Warner-Lieberman Climate Security Act would hurt the poor, raise taxes by a trillion dollars, and do next to nothing to slow climate change. It is, wrote Inhofe, "[T]he largest tax increase in U.S. history and the biggest pork bill ever contemplated with trillions of dollars in giveaways. Well-heeled lobbyists are already plotting how to divide up the federal largesse."[15] Visitors to Frontiers of Freedom's website can also find a diatribe entitled, "Climate Control: A Costly Proposal," by Heritage Foundation vice president for marketing and communication Rebecca Hagelin. She summarized the Climate Security Act of 2007 thus: "Lawmakers have cooked up an expensive solution to a hyped-up rallying cry against a "problem" that scientists can't even agree exists in the first place. Of course, Congress is doing what Congress seems to do best—pass laws in response to the latest craze."

The website, www.ff.org, also features a letter by Frontiers of Freedom Institute President George Landrith (and co-signers) in which Landrith protests mandatory limits on carbon emissions: "It is becoming increasingly apparent that the Kyoto process, with its built-in momentum for ever more unrealistic emission reduction targets, is economically ruinous and, hence, politically unsustainable."[16] The letter's co-signers include representatives of the American Conservative Union, the Competitive Enterprise Institute, and the National Center for Policy Analysis. Frontiers of Freedom also had an interview on its website in which President Landrith talks with bombastic TV commentator Glenn Beck.

In the introduction to the interview, Beck ridicules former Vice President Al Gore's book, *An Inconvenient Truth,* for using a computer-generated shot of an Antarctic ice shelf. "That's what I said about the movie," Beck quipped, "nothing real in there."

"Al Gore doesn't have a science degree," Beck continued. "He doesn't have any credibility quite frankly, either."

Beck goes on to discuss the Warner-Lieberman Climate Security Act, and Landrith makes no effort to correct Beck's unusual views. Instead, Landrith claims in the interview that by 2030, as a result of the bill, the average American family would be paying "a stealth tax" of "$6,752 a year." The following brief excerpts convey the tenor of the interview:

"**Beck** [characterizing the cap-and-trade feature of the bill]: You buy your pollution.

Landrith: Right, exactly. . . .

Beck [sarcastically]: So in other words, if we pay for our sins, that's okay.

Landrith: Exactly. And that's what's going to drive all this. That's what's going to make it more expensive, so you and I will have the privilege of paying all this extra money so that someone in Washington can feel good about themselves and feel like they did something good for the country. But it's not going to do anything to change the climate. . . .

Beck [ending the interview]: I'll tell you what this [bill] is all about. The oil companies are for it because it's going to choke the bat snot out of coal. And we're going to need the coal."

If that wasn't simplistic enough, the ideological gloves really come off on the Frontiers of Freedom website in a piece by self-described "staunch conservative" Christopher Adamo whose bio noted that he "saw, to his dismay, the nation's moral foundations being destroyed before his very eyes." In "Czech President's 'Inconvenient Challenge' to Al Gore," Adamo refers to "the theories of global warming" as "merely open-ended speculation."[17] Seizing on a debate challenge issued by

Czech President Václav Klaus to Gore, Adamo uses the news as an opportunity to attack the Nobel Prize winner for his efforts "to foment a global warming panic." Adamo reviles Gore's film, "An Inconvenient Truth," as "an insipid and unsubstantiated piece of propaganda." According to Adamo, the Oscar-winning documentary "is replete with fantastic prophesies of doom for the planet unless America immediately regresses to third-world squalor." Before the mudslinging ends, Adamo accuses Gore of "hysterical claims" and accuses liberals of exploiting fears about global warming for their own political advantage. His message is clear: if the arrogant liberals are left unchecked, they will undermine everyone's basic freedom and well-being. (This deeply misguided article was still on the website in 2012.) Not surprisingly, Frontiers of Freedom/Center for Science and Public Policy received $612,000 from ExxonMobil from 2000 to 2003.[18]

- **The George C. Marshall Institute**, another libertarian think tank, is closely identified with the anti-climate change lobby and is funded by right-wing foundations. Its former chairman and co-founder, the late Dr. Frederick Seitz, acknowledged (prior to 2000) that the Institute did not do original climate research and that its reports were fundamentally expressions of opinion. Some of the Institute's publicity work is devoted to covering the efforts of those still producing critiques of climate science, such as a Fraser Institute report titled, "The Science Isn't Settled—The Limitations of Climate Models."[19] Another publication attacks the climate models used to forecast climate change. "Current climate models have many shortcomings. . . . No climate model has been scientifically validated."[20]

- **People for the American West!**, funded by mining interests, portrayed itself during the 1990s as a grassroots group while it lobbied for mining, timber, and other industries that exploit public lands. The group appears to no longer be active.

- **The Science and Environmental Policy Project** (SEPP) was run by climate skeptic Dr. S. Fred Singer, who has taken consulting fees from at least half a dozen major oil companies as well as from cigarette and chemical firms. He once also reportedly opposed the idea that human activity had produced a hole in the ozone layer protecting the Earth from ultraviolent radiation. Dr. Singer is a professor at George Mason University and one of the nation's most vocal critics of mainstream climate science. Dr. Frederick Seitz (see page 12), another extreme skeptic, once chaired Singer's board of directors at SEPP.

  Despite the mountain of scientific evidence to the contrary from thousands of research studies, Dr. Singer denies that humans have caused climate change or that global warming is a problem. Unlike the distinguished scientists whose work he challenges, few of Dr. Singer's writings on climate science appear to have been published in respected climate science journals. Dr. Singer instead often delivers his extreme opinions to popular media, or in lectures at climate skeptic conferences, or papers for climate skeptic groups. In a recent monograph for The Heartland Institute (profiled later on pages 26-28), which SEPP characterized as a refutation of the multivolume Intergovernmental Panel on Climate Change's *Fourth Assessment Report*, Dr. Singer cites his own work fourteen times, but only two of those publications were in a refereed journal within the previous ten years—and one of those consisted of comments Dr. Singer wrote in a letter to a climate scientist.

Singer has made questionable statements about the role of cosmic rays in climate change and about the role in climate change of variations in the intensity of solar energy reaching the Earth. Singer's unorthodox climate ideas, however, are widely cited by conservative think tanks, although these think tanks make selective and misleading use of refereed climate science. With such tactics, plus his credentials as an atmospheric physicist, Dr. Singer has managed to appear often in the media with his reassuring notion that global warming is nothing to worry about. Despite his lack of credibility with most climate scientists, he nonetheless has attracted a wide following among right-wing ideologues and people swayed by his use of scientific arguments coupled with impressive-looking maps, charts, and graphs, similar to those used by mainstream climate scientists.

On the SEPP website, the organization provided Singer's peculiar version of climate science in its answers to "Frequently Asked Questions."[21]

Here are a few samples:

Question: "But isn't there climate warming already because of the increased burning of fossil fuels—oil, gas, and coal—that creates more carbon dioxide in the atmosphere?"

Answer: "True, carbon dioxide levels are rising, but the climate seems not to be warming as a result."

Follow-up question: "And why hasn't climate warmed, when theory clearly expects this to happen?"

Answer: "The answer must be that even our best current models of the atmosphere are incomplete and leave out important features."

Regarding concerns about the spread of tropical diseases in a hotter world, SEPP offers this reply: "Well, since the climate is not warming significantly, there is no immediate reason for concern."

In answer to, "Would a global warming be good or bad?" SEPP states: "Probably both, but warming is definitely better than cooling."

SEPP reprises that theme in its response to the last of its slanted FAQs: "Should we ruin our economies and cause tremendous hardship for people to counter a phantom threat?" After brushing off climate warming as "far away and a minor problem at that," SEPP points readers to its version of a bigger problem: "The near certainty of a coming ice age. Geologists tell us that the present interglacial warm period will soon come to an end. Perhaps greenhouse warming can save us from an icy fate."

While it might seem as if SEPP would be widely ignored or derided for its extreme views, by its own tally, it has been cited hundreds of times in major news media and in articles and editorials by affiliated scientists that have appeared in many of the country's leading newspapers, including the *Wall Street Journal, Miami Herald, Detroit News, Chicago Tribune* and so on. Moreover, as noted, Dr. Singer has lectured widely and been interviewed extensively in the media, his voice amplified by think tanks, websites, and conservative media. Regrettably, unwary students, policymakers, the media, and the public are likely to fall prey to the organization's self-styled brand of "sound scientific information." SEPP even prides itself on its website in having advised the late novelist Michael Crichton on his anti-environmental thriller, *State of Fear*—in which environmentalists team up with ecoterrorists to create unjustified fear about a global warming catastrophe.

- **The National Center for Policy Analysis** sent out a 2008 press release in opposition to the Lieberman-Warner Climate Security Act of 2007 as the bill came up for Senate debate. Leading with the scary headline, "Greenhouse Gas Bill Could Raise Gas Prices to $8/ Gallon,"[22] the press release contained allegations from the National Association of Manufacturers claiming that, if the legislation passed, "the economic hit to Texas alone could force companies to cut as many as 335,000 jobs to pay for added costs." The release then went on to quote "NCPA Expert" Senior Fellow H. Sterling Burnett, who raised the specter of global cooling. The release ends with: "How can proponents fight warming when there is no warming going on?" Unwitting media who received the release might have assumed that the organization was an objective and independent policy group since it describes itself as, "an internationally known nonprofit, nonpartisan research institute with offices in Dallas and Washington, D.C. that advocates private solutions to public policy problems." No mention of the $205,000 that the organization allegedly received from ExxonMobil just between 2000 and 2003,[23] or of its affiliation with climate change denier Dr. S. Fred Singer.[24] (For more about Dr. Singer, see pages 22-25 and 29).

- **The Heartland Institute** is a loyal member of the climate science denial camp that downplays the significance of climate change and tries to cast doubt on climate science. The Institute has prominently displayed posters on its website proclaiming, "CLIMATE CHANGE IS NOT A CRISIS," and another with the legend, "Global Warming: Not a Crisis."[25] The Institute says its mission is to discover, develop, and promote free-market solutions to social and economic problems. Its Senior Fellow for Climate Change, agricultural specialist Dennis Avery, wrote *Unstoppable Global Warming—Every 1,500 Years* with leading climate skeptic Dr. Singer (above). At a Heartland Institute international conference of climate science deniers and skeptics in May 2008, Singer "rocked the crowd," in the words of Heartland President Joseph Bast, when he announced the release of a report that Bast called, "an authoritative rebuttal of the Intergovernmental Panel on Climate Change's *Fourth Assessment Report*."[26] The document Bast referred to turned out to be a paper of about 25 pages, edited by Singer, which recycled long-discredited claims about alleged Intergovernmental Panel on Climate Change (IPCC) measurement errors. The report claimed, "human greenhouse gas contribution to current warming is insignificant" and "increasing carbon dioxide is not responsible for current warming." Therefore, the report held—as all deniers of climate science would have us believe—that "policies adopted and called for in the name of 'fighting global warming' are unnecessary."[27] The Heartland Institute received $312,500 from ExxonMobil between 2000 and 2003.[28]

In early 2012, a Heartland Institute fundraising plan sent to the press by a climate scientist revealed the institute had a 2011 budget of $4.6 million, nearly two-thirds of which was allocated to global warming projects, and that it planned to raise over $7 million in 2012. An Institute staffer acknowledged that the leaked fundraising plan and donor list were genuine, stating, "The Heartland Institute apologizes to the donors whose identities were revealed by this theft."[29]

The plan showed that most of the Institute's funding came from large donors giving over $10,000 and that a single anonymous donor had contributed up to half of the Institute's budget, including more than $1.25 million annually from 2005 to 2010. The Charles G. Koch Foundation was listed among the Institute's 2011 donors.

Other donors listed for 2011 included Microsoft Corporation, Time Warner Cable, Verizon, State Farm, Nucor Corporation, Reynolds American, General Motors Foundation, GlaxoSmithKline, Comcast Corporation, AT&T, Allied World Assurance Company Holdings, Ltd., Altria Client Services Inc.,[a] and Anheuser-Busch Companies Inc. Donors listed for 2010 included Amgen, Eli Lilly & Company, Pfizer, Pepsico, and many of the corporations who continued supporting Heartland in 2011.

Employing the climate deniers' strategy of casting doubt where genuine scientific objections are lacking, the plan provides details about the hiring of a consultant to produce curriculum on climate change for vulnerable K-12 students to create the impression that human-induced climate change is a "major scientific controversy" and that climate models are controversial.

- **The Heritage Foundation** is one of the nation's largest, wealthiest, and most conservative think tanks. The organization has influenced public discourse and perceptions on the climate issue by doing its own less than impartial economic and policy analysis, which it then circulates

[a] Altria Group is the parent company of Philip Morris USA and other entities; the Altria Client Services firm engages in lobbying.

widely through the media. Here's how the foundation attacked the failed Lieberman-Warner Climate Security Act: "The Heritage Foundation concludes that [the] economic consequences [of the Lieberman-Warner Climate Security Act] will be dire. In the manufacturing sector nearly 3 million blue-collar jobs would disappear. Incomes would fall, fuel prices would skyrocket, and carbon tariffs on goods imported from countries that do not meet the bill's rigorous standards would add to the cost of imported clothes, consumer electronics, and automobiles."[30] Report author Michael Franc is Heritage Foundation's Vice President for Government Relations and was listed on the organization's website as an expert on health care.[31] His article also included a strong pitch for nuclear energy, a technology apparently favored by the foundation: "There are only two ways to achieve Lieberman-Warner's objective without flat-lining the economy," Franc writes. "The first is by increasing dramatically our use of clean nuclear energy." The other way, according to Franc, is carbon capture and sequestration.

The charges about the climate bill's disastrous economic consequences evidently originated in an in-house Heritage Foundation report called, "The Economic Costs of the Lieberman-Warner Climate Change Legislation."[32]

Heritage reportedly received $680,000 from ExxonMobil from 1998 to 2010,[33,34] and a total of $3.35 million from Koch Industries, whose business includes petroleum refining and pipelines. (Koch Industries has also supported the American Enterprise Institute and the Cato Institute.)[35]

- **The Hoover Institution**, a large and wealthy bastion of conservative thought at Stanford University, has endorsed the views of climate deniers. One 2008 Hoover Institution Essay in Public Policy, for example, reads, "The scientific evidence for a presumed 'human influence' [on global climate] is spurious and based mostly on the selective use of data and choice of particular time periods."[36] The author goes on to provide his reasons why "global warming (GW), if it were to take place, is generally beneficial. . ." The author again turns out to be none other than the ubiquitous Dr. S. Fred Singer. Some of the titles on the institution's website indicate how seriously the organization takes concerns about global warming: "Global Warming and Globaloney," "The Pseudoscience of Global Warming," "Global warming causes everything, Part II," and "Will Global Warming Save Lives?" Among the Institution's many sources of funds, it received $295,000 since 1998 from ExxonMobil.[37]

In "Put a Tiger in Your Think Tank, Part II" (cited earlier in this chapter), the authors list some 40 organizations that received ExxonMobil funding between 2000 and 2003. Many have devoted much of their energy to raising unfounded doubts about global warming and characterizing global warming as a controversial, speculative scientific theory based on tenuous computer models.

As noted, some of these groups have also flatly denied that global warming exists. Others ridicule the conclusions of climate science. The roster of skeptic organizations covered in "Put a Tiger in Your Think Tank, Part II" includes: Acton Institute for the Study of Religious Liberty, Advancement of Sound Science Center, American Council for Capital Formation, American Council on Science and Health, American Enterprise Institute, American Legislative Exchange Council, Annapolis Center for Science–Based Public

Policy, Atlas Economic Research Foundation, Cato Institute, Capital Research Center, Center for the New Europe, Center for the Defense of Free Enterprise, Center for the Study of Carbon Dioxide and Global Change, Citizens for a Sound Economy, Committee for a Constructive Tomorrow, Congress of Racial Equality, Consumer Alert, Federalist Society for Law and Public Policy Studies, Foundation for Research on Economics and Environment, Fraser Institute, Free Enterprise Action Institute, George C. Marshall Institute, Heartland Institute, Hudson Institute, Independent Institute, Institute for Energy Research, International Policy Network, Mackinac Center for Public Policy, Media Research Center, Mercatus Center, National Black Chamber of Commerce, National Center for Policy Analysis, National Center for Public Policy Research, Pacific Legal Foundation, Pacific Research Institute for Public Policy, Property and Environment Research Center, Reason Foundation, Science and Environmental Policy Project, and Tech Central Science Foundation.

Collectively, the conservative and libertarian think tanks above have exerted a profound effect on the climate debate. These ideological allies of the fossil fuel industry and other large corporations often tend toward anti-environmentalism in general and have encouraged an unfounded skepticism about climate change in particular.[38]

POLITICAL OPPONENTS DELAY MAJOR CLIMATE LEGISLATION

Regrettably, the Warner-Lieberman Climate Security Act was killed in the U.S. Senate in 2008 by predominantly Republican opponents who insisted that all 492 pages of the bill be read aloud on the Senate floor. The process took some nine hours and supporters of the bill were unable to garner 60 votes needed to bring debate on the measure to a close so a vote could be taken, even though a majority of senators had given the bill their support. If the bill had passed, it would have needed 60 votes to override the veto that President George W. Bush promised to exercise if it ever reached his desk. Thus, the Senate failed in 2008 to adopt climate legislation designed to cut U.S. carbon emissions 70 percent by 2050 through the use of a cap-and-trade system.[39] The not-so-subtle lobbying by a number of the "objective," "nonpartisan," "nonprofit" organizations described above undoubtedly contributed greatly to that result.

More Moderate Business Coalitions on Climate Policy

Not all fossil fuel producers, nor all large corporate fossil fuel consumers, chose to cast their lots in with the Global Climate Coalition and its questionable friends during the 1990s. Various large oil, gas, chemical, and utility companies gradually joined the Business Environmental Leadership Council (BELC) of the Pew Center on Global Climate Change, distancing themselves from fellow industry members who categorically opposed carbon emission reductions.

BELC members included the American Electric Power Company, BP, CH2M-Hill, Dupont, the now-defunct Enron Corp., Shell International, Sunoco, and Toyota. These companies, circa 1999, played a leadership role in adopting positions on climate change that were not initially popular in their industries.

BELC accepts the views of most scientists that enough is known about climate change to take action. It advocates emission reductions by business through improved energy efficiency, and it recognizes the Kyoto agreement as an important first step toward further international cooperation.

Some of the companies in this camp made important investments in energy efficiency and a few—BP (then known as BP-Amoco) and Shell—invested significantly in clean energy technologies as well, although those commitments were small relative to the size of the companies' fossil fuel businesses. To their credit, BP, Shell, and Elf Aquitaine committed themselves to reducing their own emissions below the targets of the Kyoto Protocol, and both BP and Shell publicly acknowledged that an eventual shift to noncarbon fuels is inevitable.

These companies seemed prepared by 2000 to accept a slow global movement toward clean energy technologies (and were positioning themselves to profit from it), so long as it was gradual enough not to jeopardize their fossil fuel revenues. Whereas the stand of the BELC firms took some foresight and pluck, it left them far from a vanguard position on how urgently the world should pursue a transition to clean energy. The BELC companies mainly supported neoliberal market mechanisms for solving climate problems, including heavy reliance on emissions trading and voluntary emission reductions.

The Business Council for a Sustainable Energy Future is another business coalition that supports moderate actions to stem climate change. Members include companies engaged in renewable energy, energy efficiency, natural gas, and a few electric utilities. (The natural gas industry thrives on concern about carbon dioxide and sees it as a way of taking market share from the coal industry.) It now also includes power developers, equipment manufacturers, independent power generators, green power marketers, retailers, gas and electric utilities.[b]

Whereas the moderates on climate issues in the business community are not as likely to stand in the way of measures to combat climate change, they are also unlikely to be the driving force that produces a timely solution. They tend to agree with the idea of carbon emissions trading but have often balked at the notion of a carbon tax.

---

[b] www.bcse.org/index.php?option=com_content&task=view&id=13&Itemid=39.

Chapter 3

Prominent Climate Myths and What Science Says About Them

Let's analyze the principal myths that the hard-line oil and chemical company interests have used to build their case about climate change. Although these myths sound plausible and appealing to people unfamiliar with climate issues, science has conclusively refuted them.

To understand why these myths are fabrications, one needs to be familiar with some basic concepts in climate science. I will therefore briefly explain how global heating and cooling are affected by the presence of airborne global heating gases. I will also explain how we can tell which gases are most important and how they relate to each other in terms of their strength and persistence in the atmosphere. In the course of the discussion, I'll introduce and explain a few concepts like global warming potential, radiative forcing, and atmospheric residence time.

* * *

Myth One

The scientific foundation for concerns about climate change is uncertain and unproven. The evidence is contradictory and inconclusive.

Response:

After years of research—and the preparation of thousands of meticulously peer-reviewed studies—some 2,500 of the world's leading scientists with a deep knowledge of climate issues reached agreement on the causes of climate change and its probable impacts. Their work was conducted by the Intergovernmental Panel on Climate Change (IPCC), a multinational, multidisciplinary research alliance operating under the auspices of the U.N. and the World Meteorological Organization. A more thoughtful, knowledgeable, diligent, impeccably credentialed, and cautious scientific body can hardly be imagined. Its scientific work has been collected in a series of massive climate change Assessment Reports available to all at www.ipcc.ch.

These landmark Assessment Reports have appeared at five- or six-year intervals since 1990 (1990, 1995, 2001, and 2007); the next one is due by 2014. In each, the conclusions have become more stark and unambiguous.

In the First Assessment Report, the scientists concluded that they could not yet unequivocally attribute rising global temperatures to human activity. The Second Assessment Report, however, declared in a cautiously framed conclusion that humanity was indeed changing the climate.

"The balance of evidence," they wrote, "suggests a *discernible human influence on global climate*." These carefully chosen words drew worldwide attention and were widely quoted, though little global action was taken to diminish the threat to which the report pointed.

Since that subdued finding in 1995, the panel's language has grown progressively more forthright. The Third Assessment Report stated, "Most of the observed warming over the last 50 years *is likely to have been due to the increase in greenhouse gas concentrations*."

In the most recent Fourth Assessment Report, the panel stated: "Most of the observed increase in global average temperatures since the mid-20th century *is very likely due to the observed increase in anthropogenic [man-made] greenhouse gas concentrations*." The report also declared: "*Warming of the climate system is unequivocal*, as is now evident from observations of increases in global average air and ocean temperatures, widespread melting of snow and ice, and rising global average sea level." The forthcoming Fifth Assessment Report is likely to use even bolder, more unqualified language.

The IPCC Assessment Reports and the climate change forecasts they contain, which have become increasingly dire, integrate a huge body of long-term physical data gathered from all over the world by experts in many scientific disciplines. Their comprehensive and evenhanded studies have become the standard reference for policymakers, scientists, and all serious students of climate issues.

With each new multiyear assessment, the respected alliance has expressed greater and greater levels of scientific certainty about its fundamental conclusion that human activity is rapidly warming the climate and that the perturbations are threatening millions of species with extinction and endangering the survival of life as we know it on Earth.

The IPCC scientists' detailed knowledge of how climate operates is deduced from fundamental principles of physics, chemistry, biology, ecology, geology, geography, oceanography, and atmospheric sciences. These findings are all supported by vast amounts of data gathered from all over the world and cross-checked in meticulous peer-reviewed processes. All these data and scientific insight are then used in the creation of highly sophisticated computer models of climate processes to aid in their analyses.

The IPCC's most recent Fourth Assessment Report reflected the efforts of more than 600 scientists in 41 nations. Despite their broad scientific agreement, and the millions of hours of labor invested in attaining it, the scientists nonetheless were fair-minded enough to prominently identify a number of uncertainties in their findings.

The uncertainties largely pertained to the exact rate of climate change expected, the magnitude of its impacts, and their regional effects. As the years have passed since the First

Assessment Report, the scientific uncertainties about climate change have diminished. Moreover, the IPCC's forecasts have proven to be conservative rather than alarmist relative to the degree of warming actually observed, compared with the warming forecast.

Yet given the tremendous complexity of the systems being studied, it is hardly surprising that precise answers were not, and are still not, available to all conceivable questions. Whereas the remaining uncertainties cannot be dismissed as trivial, they are of far less importance than the overarching conclusion that human actions are disrupting the climate and that, if the disruptive behavior is not corrected, the world will inevitably suffer a worsening series of tragic consequences.

Climate scientists may have inadvertently fostered the notion that the uncertainties remaining are more significant and fundamental than they really are because true scientists tend to take their areas of agreement for granted and focus on their remaining differences, however minor, in the spirit of discovering the truth.

"We don't know everything about this subject but we know a lot," says Professor John P. Holdren of Harvard, President Obama's science advisor. "And what we know suggests that the downside risks of failing to deal with it are very large."

So should we wait until all uncertainties are resolved and all answers are in? The late Stephen Schneider, a distinguished member of the Stanford University faculty and a member of the IPCC, disagreed. "Science," he observed, "is never 100 percent certain of anything." To hold action hostage to a standard of 100 percent certainty would be absurd and, in this case, dangerous.

* * *

Myth Two

Even if humans added substantially to the atmosphere's carbon dioxide concentration, carbon dioxide is not a powerful

enough greenhouse gas to cause global warming. Other gases such as nitrous oxide, methane, chlorofluorocarbons, and even water vapor are far more powerful.

Response:

In the 1820s, the renowned French mathematician and physicist Joseph Fourier first explained that when visible light from the sun strikes the Earth and is transformed to heat, atmospheric gases trap the resulting heat radiated skyward by the Earth, thereby warming the planet. His insight was the first general description of what today is called "the greenhouse effect." The theory was elaborated and formulated mathematically in 1896 by the great Swedish chemist Svante Arrhenius.[1]

He focused his attention on carbon dioxide and water vapor and deduced that adding carbon dioxide to the atmosphere will cause global heating, while reducing it sufficiently would send the Earth into an ice age.[2] He was thus the first person to understand that the injection of carbon into the atmosphere by the continued burning of fossil fuel would heat the Earth. His brilliant 19th century calculation that doubling the atmosphere's carbon dioxide content would raise the Earth's temperature by a few degrees is remarkably close to contemporary climate sensitivity estimates by IPCC scientists.[a]

Although other gases are more powerful per molecule at trapping solar radiation than carbon dioxide, the overall effect of a gas on the heat balance of the Earth is a function of its potency per molecule as a global heating gas multiplied by its mass in the atmosphere. The term used to describe this effect on the net change in incoming and outgoing radiation to the planet caused by a gas is its "radiative forcing."

---

[a] Arrhenius predicted that doubling atmospheric carbon would raise the Earth's temperature about 7 ° F. The 2007 IPCC report predicted a temperature rise of 3.3-7.5 ° F.

Given the mass of carbon dioxide in the atmosphere and its per molecule potency, it has a greater total radiative forcing[b] than any of the other global heating gases except water vapor. Therefore, carbon dioxide has a greater effect on atmospheric temperature than any global heating gas (except for water vapor). The effect of a gas on the Earth's heat balance is averaged over the Earth's entire surface, which we can measure in square meters (m^2), so the radiative forcing of each global heating gas is measured in units of Watts per square meter (W/m^2).

Whereas water vapor has the greatest heating effect of all the global heating gases, humans do not have an important *direct* effect on the amount of water vapor. Almost all the water vapor in the atmosphere is produced by evaporation from land and water surfaces and by a process called evapotranspiration. (During evapotranspiration, plants draw water from the soil through their roots and release it as vapor to the atmosphere through their leaves.) As we will see, however, humans do affect the amount of water vapor and clouds *indirectly* in an extremely important way, through an effect on the atmosphere's global heating gas concentration.

Although other gases are more powerful heat trappers than carbon dioxide on a per molecule basis, the amount of carbon dioxide in the atmosphere—measured in billions of tons—is so much greater than the mass of the other important greenhouse gases, such as methane, nitrous oxide, chlorofluorocarbons, and ozone, that global heating is strongly dominated by the concentration of carbon dioxide.

The concentration of carbon dioxide in the atmosphere is measured in parts per million (ppm) by volume, whereas the concentration of methane, nitrous oxide, and ozone are only present at parts per billion (a thousand times less) and

---

[b] The Earth both receives energy from the Sun and reradiates energy back into space. Radiative forcing is used in this book simply to mean the difference in the Earth's energy balance produced by the effect of greenhouse gases and aerosols on incoming and outgoing radiation and hence on Earth's temperature.

chlorofluorocarbons are only found in concentrations of parts per trillion.

Thus, based both on atmospheric concentration and intrinsic potency at trapping heat, scientists have calculated that the radiative forcing of carbon dioxide is 1.53 W/m^2. Methane's radiative forcing is 0.48 W/m^2. Nitrous oxide's radiative forcing is 0.15 W/m^2. Carbon dioxide's radiative force is thus more than three times that of methane and more than ten times that of nitrous oxide.[3] This is why its atmospheric concentration is of such concern.

But the effect of a heat-trapping gas on the atmosphere's temperature (and thus indirectly on the Earth's temperature) is not only determined by its per molecule heat absorption and its concentration in the atmosphere, but by how long it stays in the atmosphere, called its atmospheric "residence time." Water vapor, for example, though powerful in trapping heat because of its per molecule potency and enormous volume in the atmosphere, nonetheless has an extremely short residence time of only 9 days. By contrast, carbon dioxide has a long residence time: a portion of the carbon dioxide added to the atmosphere every moment remains there for thousands of years, although some is removed much more rapidly.[c]

Thus, among the global heating gases, carbon dioxide's predominance—second to water vapor—and its longevity in the atmosphere more than compensate for the greater potency on a per molecule basis of methane and nitrous oxide. Therefore, the resultant heating from carbon dioxide is far greater than that of

[c] The fact that water vapor has such a short residence time in the atmosphere should not be misunderstood as an indication of its lack of potency as a global heating gas. Water vapor and clouds, a form of water vapor, account for far more than half the global heating initiated by heat-trapping gases. They do so in the context of climate change, however, in a feedback response to the higher average surface temperatures initiated by the increase in the atmospheric concentration of global heating gases. Clouds both cool the Earth by reflecting solar energy back into space, and they warm the Earth by intercepting and reradiating outgoing heat. On balance, they tend to cool the Earth. See "Cloud forcing" at www.en.wikipedia.org/wiki/Cloud_forcing for more information.

methane and nitrous oxide combined. The global warming potential of a global heating gas is its radiative forcing multiplied by the years that the gas remains in the atmosphere before removal by natural processes.[4]

As the amount of carbon dioxide increases, it raises global temperature, as explained earlier (pages 39-41). Because warmer air can hold more moisture than colder air, the moisture content of the atmosphere increases with temperature. Thus, as the carbon dioxide concentration increases and raises atmospheric temperature, more water will evaporate and further amplify the heating induced by the carbon dioxide alone. This secondary "feedback effect" thus magnifies the heating effect initially produced by an increase in carbon dioxide and is accounted for in the standard value used by scientists for carbon dioxide's global warming potential. That standard value has been given as 1 and is used as a reference value for comparing the global heating potential of other gases to carbon dioxide. The global warming potential of other greenhouse gases is given as multiples of that value.

* * *

Myth Three

Human burning of fossil fuels is not the source of observed increases in atmospheric carbon dioxide concentration above naturally occurring levels. Increases are caused by natural processes, such as the outgassing of the Earth's mantle.

Response:

Carbon in the form of carbon dioxide is released from the Earth's mantle into the atmosphere through volcanic hot spots, volcanic subduction zones (places where the great plates of the Earth's crust dive beneath each other), and through midocean ridges.

The element carbon has three naturally occurring isotopes. Isotopes are related forms of the same element. They behave the same chemically and differ only in mass. The isotopes of carbon are carbon-12, carbon-13, and carbon-14. They occur in nature in known ratios.

Carbon-12 is by far the most common, accounting for almost 99 percent of all the carbon on Earth. Most of the other 1 percent of the carbon is carbon-13. Unlike carbon-12 and carbon-13, which are stable, carbon-14 is radioactive and extremely rare in nature.

Plants preferentially utilize the lighter carbon-12, and so carbon derived from fossil fuels—which are largely formed from dead, decayed, and compressed ancient plant matter—have a higher carbon-12 to carbon-13 ratio than carbon from other sources. Fossil fuels—being derived from the decayed remains of ancient plants—have the same carbon isotope ratio as plants. Thus the increase in this ratio that has been detected in atmospheric carbon dioxide samples over time confirms that the carbon dioxide increases of concern are coming from the burning of fossil fuels and other plant material, rather than from an increase in outgassing from the Earth's mantle.

Although the outgassing of the Earth's mantle has released vast quantities of carbon dioxide since the planet's formation billions of years ago, the outgassed carbon dioxide does not exhibit an enrichment in carbon-12. It therefore cannot account for the enrichment of carbon-12 found in the atmosphere.

The outgassing of the Earth's mantle as it relates to climate and atmospheric chemistry is explained much more fully in the *Intergovernmental Panel on Climate Change's Working Group I Report* and is also well explained on the award-winning science website, www.Realclimate.org.[5]

Yet another type of information supports the conclusion that the recent increase in the atmosphere's carbon dioxide content is being caused by the burning of fossil fuels and the destruction of forests. It is corroborated by data on how much fossil fuel has been burned and how many tons of carbon-containing forests have been destroyed.

The tonnage of carbon dioxide in the atmosphere corresponds well to the quantity of carbon dioxide generated by the burning of fossil fuels and forests (some 500 billion metric tons) minus the Earth's well-understood natural carbon dioxide removal processes,[6] just as it clearly does not correspond to any outgassing of carbon dioxide from the Earth's mantle.

To sum up, based on reasonable scientific approximations of the quantity of carbon dioxide added and removed from the atmosphere, it's possible to calculate the concentration of carbon that would be expected in the atmosphere if the source of the increase were the burning of fossil fuels and clearance of forest. Sure enough, these estimates are in close agreement with the actual concentration of carbon dioxide in the atmosphere, which as of mid-2012 was 394 ppm.[7]

* * *

Myth Four

Climate varies naturally. We are in a natural warming cycle that has little or nothing to do with human influence. There have been warm periods in the past, such as the Medieval Warm Period, that prove the world heats and cools naturally, unaffected by greenhouse gas emissions from human industrial activity. Indeed, periods of warming may actually be caused by natural fluctuations in cosmic rays or solar radiation.

Response:

From 1900 to 2000, the world's temperature rose about 1° F—*ten times faster* than the rate of temperature increase characteristic of warming periods over past millennia. That strongly suggests that the climate change underway is *not* natural. The Earth's temperature is now about 1.4° F hotter than at the start of the Industrial Revolution beginning about 1750.

In addition, the recent rise in global temperatures bears a "human fingerprint." Climate scientists have identified a complex but predictable pattern that accompanies atmospheric warming caused by human action. This phenomenon, which involves cooling of the upper atmosphere and warming of the lower atmosphere, is characteristic of the global temperature changes now observed. The probability that natural fluctuations could reproduce these consistent patterns by chance, without human intervention, is extremely low.

Climate science deniers also claim that natural variations in solar radiation are responsible for changes in global temperature. But studies based on sophisticated analysis of several centuries of temperature data show that the changes in solar energy could account for no more than a tenth of the global warming observed over the past century. Furthermore, climate models indicate that current natural factors, such as changes in the Earth's orbit or the Earth's tilt in space, are likely to be responsible for a gradual *cooling* of the Earth, starting 6,000 years ago. The warming of the past 100 years, therefore, is especially unnatural in that it interrupted, and then very dramatically reversed, a prolonged natural cooling cycle.

As already explained (see response to Myth 2), the increase in carbon dioxide now found in the atmosphere does not come from increased volcanism. Another bogus assertion is that the carbon dioxide warming the atmosphere is being released at an unusual accelerated rate by the oceans, not by human activity. Data on the carbon concentration in the ocean, however, reveal this assertion to be false.

The carbon concentration in the oceans it turns out, is increasing, not decreasing. The increase, in fact, is directly measurable and has produced an increase in the amount of dissolved carbonic acid. That in turn is making the oceans more acidic, threatening the well-being of marine organisms with shells and other calcium-based external skeletons, like coral's, that dissolve in acid.

The existence of a Medieval Warm Period that supposedly was pervasively warmer than today's climate is in serious dispute. It has not been proven that the Medieval Warm Period was more than a regional or at best a hemispheric phenomenon, rather than a true global temperature anomaly.[d] Regional warming and cooling anomalies are accounted for by climate science and are not inconsistent with overall trends in global warming. Secondly, data about the Medieval Warm Period have been shown to contain errors, leading to questions about whether the Medieval Warm Period was indeed significantly warmer than a decade ago. Recent temperatures have actually surpassed those of the Medieval Warm Period.

The notion that cosmic rays account for the global warming trend being observed has no basis in fact. The intensity of cosmic rays reaching the Earth is difficult to measure, but no consistent trends in cosmic ray intensity have been correlated with global temperature. By contrast, the concentrations of carbon dioxide, methane, and nitrous oxide *have* correlated closely with global temperature cycles for more than 700,000 years, and scientific data, especially the careful analysis of ancient ice cores with trapped air bubbles, has conclusively proven this.

* * *

Myth Five
The world is really cooling, not warming.

Response:

An enormous collection of data consisting of millions of temperature readings taken all over the land, sky, and sea conclusively show that the Earth is warming. In addition, since 1880, nine of the ten warmest years have occurred between 2000 and 2012.[8] The proposition that the world is cooling would

[d] www.realclimate.org/index.php/archives/2004/11/medieval-warm-period-mwp/ and www.ncdc.noaa.gov/paleo/globalwarming/medieval.html.

require that the entire global community of scientific climate experts have been totally blind-sided on the basic direction of global temperature change.

The climate science deniers making this argument long ago ran out of legitimate objections to climate science and therefore have relied on claims like:

- Global temperature readings are upwardly biased because they come from weather stations in or around cities where the "heat island" effect has distorted temperature readings. (The "heat island" effect is the tendency of urban areas to be significantly warmer than their surroundings mainly because of the heat-retention properties of the brick, asphalt, and cement surfaces and, secondarily, because of heat from fossil fuel combustion.)[9]
- Temperature measurements were made on land, not over cooler ocean surfaces.
- Even if surface temperatures have gone up, satellite data prove that the Earth is cooling.
- The Earth cooled from 1940 to 1970, despite a global increase in the rate of carbon dioxide emissions.

The first two objections are simply false. Urban temperature data have been carefully corrected for urban heat island effects, which in any case are minimal,[e] and ample temperature

---

[e] Climate researchers at Stanford University using high-resolution measurements of atmospheric variables concluded that the heat island effect could account for no more than 24 percent of gross global warming. See www.stanford.edu/group/efmh/jacobson/Articles/Others/HeatIsland+WhiteRfs0911.pdf. More recently, a massive re-analysis of global warming data by a team headed by erstwhile climate skeptic and UC Berkeley physicist Richard Mueller (funded in part by the Koch Foundation, established by the conservative billionaires Charles and David Koch) thoroughly and publicly confirmed the findings of the IPCC with regard to observed warming. Finally, measurements of the heat content of the upper layers of the world's oceans (which are not susceptible to urban heat island effects), by three independent research groups absolutely show that ocean temperatures have increased substantially since the 1970s. See www.aip.org/history/climate/20ctrend.htm#N_33_.

measurements have been taken at sea. A distinct upward temperature trend is evident in all these data.

Regarding satellite data, it is true that satellite measurements of temperatures at altitudes of 5,000–30,000 feet indicate global warming at less than the rate forecast by some climate models. However, they do not refute that the Earth is warming. First, readings taken over a thicker layer of the atmosphere from the same satellite are significantly warmer. Second, the use of satellites for temperature measurements is a relatively new development. To extract long-term trends will require a longer historical record. Finally, some remotely sensed tropical satellite data were found to incorrectly indicate cooling of the Earth because of a drift in the satellites' orbits. That drift caused the satellites to misrecord and mix cooler evening temperatures with the warmer daytime readings they were supposed to be recording, biasing their results.[f]

By contrast, the finding that the Earth is warming rests on more than a century and a half of surface temperature measurements, and on the historic climate record obtained from natural indicators going back more than 700,000 years.[10]

The argument that the scientifically acknowledged period of global cooling from 1940 to 1970 refutes a longer-term global warming trend is also false. The occurrence of several major volcanic eruptions during this period plus the coincidence of reduced solar sunspot activity resulted in a limited short-term period of cooling.

Volcanic eruptions release shiny sulfate aerosols into the upper atmosphere, where they reflect solar energy and tend to cool the Earth. Sunspot activity coincides with increased solar radiation; hence, the reduction in sunspot activity had a cooling effect by diminishing the amount of solar energy reaching the Earth. By about 1970, however, the reduction in volcanic activity, the normalization of sunspot activity, and the progressive increase in atmospheric carbon dioxide concentration due to

---

[f] www.agu.org/journals/jd/jd0706/2005JD006881.

human activity caused global temperatures to resume their dramatic and long-term upward trend.

* * *

Myth Six

There's no cause for alarm. The warming observed is gradual and slight. The dire projections of climate change impacts are vastly exaggerated.

Response:

The global scientific consensus published in 2007 by the IPCC projected a range of average Earth surface temperature increases. The projections also covered six economic and population growth scenarios. The rosiest scenario implied that global average temperature would rise by 2–5° F by the end of this century compared with late-20th century temperatures.

But if trends more closely approximate the most pessimistic economic and population trajectory, the IPCC projected a global average temperature increase of 4–11° F.[g] Some commentators may have concluded that they could simply crunch these two ranges together and determine from their overlap which temperature range would be most likely. The problem is that the IPCC did not provide the probabilities of their six different global emissions scenarios. And without definitive knowledge of which scenario the world is going to follow, it is impossible to identify the likeliest range of expected temperature change.

Currently, the world appears to more closely approximate the high-emissions scenario than the low- or mid-range scenarios. From 1990 to 2010, for example, annual carbon dioxide emissions rose an astonishing 45 percent.[h] In an even shorter period from 1992 to 2007, China's emissions increased almost

---

[g] www.epa.gov/climatechange/science/futuretc.html and IPCC. *Climate Change: 2007*, vol. 1.

[h] "Carbon Emissions Hit All-time High," *Geographical*, Royal Geographical Society, November 2011.

300 percent.[i] And just in the single year between 2009 and 2010, global emissions grew an unprecedented 5.8 percent.[j]

Although even a dramatic temperature rise of 4–11° F may at first seem small to some people, that rise is more than enough to change the incidence of extreme high and low temperatures (which affect plant and insect life cycles), alter the timing of seasons, raise sea level, and modify precipitation patterns.

Remember that ice age temperatures averaged only 9° F colder than present. A few degrees can make a huge difference. A shift in global average temperature of several degrees over a century is not gradual—it is stunningly fast on a geological timescale. (The change into and out of an ice age might naturally take on the order of 20,000 years.)[k] Because of the swiftness of contemporary warming, many ecosystems will be unable to adapt and will perish or be severely stressed.

Whatever emissions scenario the world follows and whatever the increase in its average temperature may be over the next 90 years or so, the higher latitudes of the northern hemisphere will warm substantially more than the global average because of positive feedback processes. These include warming processes involving the melting of snow and ice.[l] Some models project that temperatures over Canada and the northern half of Asia will increase four times faster than the average.[m] The interior temperature of continents will also increase faster than the average.

* * *

[i] "Construction Driving China's Emissions Growth," *Geographical*, Royal Geographical Society, November 2011.

[j] Carbon Emissions Hit All-time High," *Geographical*, Royal Geographical Society, November 2011.

[k] www.aip.org/history/climate/cycles.htm.

[l] www.earthobservatory.nasa.gov/Features/GlobalWarming/page5.php.

[m] M.G. Sanderson, D.L. Hemming, and R.A. Betts, *Regional Temperature and Precipitation Changes Under High-end Global Warming (2011)*, www.rsta.royalsocietypublishing.org/content/369/1934/85.full#fn-group-1.

Myth Seven

Climate change takes place slowly and is therefore a long-term problem that merits a long-term solution, not a quick fix. We should wait and see what happens.

Response:

Time is of the essence in keeping heat-trapping gas emissions from getting dangerously higher. A report by the European Commission's Joint Research Centre and PBL Netherlands Environmental Assessment Agency found that, as noted on the previous page, human-induced global carbon dioxide emissions grew 45 percent in only the 20 years from 1990 to 2010.[n] According to the U.S. Department of Energy's Energy Information Administration, global carbon emissions will be almost twice 1990 levels by 2035.[o] This assumes continued increases in energy use and a failure to shift significantly from fossil to renewable energy. Acting promptly to lower emissions rates now, however, will reduce risk, provide economic benefits, and mitigate damage.

Of course, it is not enough to merely halt the increase in carbon dioxide emissions. The climate cannot tolerate the stabilization of carbon emissions at current levels without producing more powerful storms, tornadoes, droughts, and floods. Not only are current emissions rates too high, but even if we were to hold current emissions steady, which would be a big improvement, the concentration of carbon dioxide and other greenhouse gases in the atmosphere would still continue rising for hundreds of years, since they're already pouring into the atmosphere far faster than natural processes can remove them.

---

[n] "Carbon Emissions Hit All-Time High," *Geographical*, Royal Geographical Society, November 2011.

[o] U.S. Department of Energy, Energy Information Administration, *International Energy Statistics*, www.eia.doe.gov. and Energy Information Administration, 2011 *Energy Outlook*, www.eia.gov/forecasts/ieo/more_highlights.cfm#world and www.eia.doe.gov/iea.

Thus, even holding emissions steady will worsen the climate crisis. Furthermore, even if the concentration of carbon dioxide could be magically lowered to a historically safe level, the Earth would still continue heating for hundreds of years because of stored heat the oceans will be gradually releasing.

It should also be clear that if the concentration of carbon dioxide is stabilized at a level that is substantially above values found during the past 10,000 years, then that new atmosphere is simply not compatible with our continuing enjoyment of the familiar climate that has endured for all of recorded human history. If the carbon dioxide level in the atmosphere is abnormal, the climate will be abnormal. A stable but elevated concentration of carbon dioxide relative to the 10,000 year norm will force the temperature to rise. It will continue to do so until it corresponds to the much hotter climate that the record of climate data from the geological past shows occurred when carbon dioxide levels were 400 ppm or above. (Think palm trees and crocodiles in the Arctic.)

For a more livable, not to mention a more pleasant, climate like the one in which humanity has existed since the last ice age, not only will emissions need to be stabilized, but they will need to be reduced below the rate at which carbon dioxide is removed from the atmosphere by natural processes. Otherwise the concentration of carbon dioxide in the atmosphere will continue to increase, even though emissions might be stable. If any reservoir is filled faster than it's drained, even if it's filled at a constant rate, the level of the contents will rise.

Currently, about half the carbon dioxide humans pump into the atmosphere is removed by natural processes in about 30 years, with about 30 percent staying in the atmosphere for a few hundred years and about 20 percent staying for thousands of years.[11] By contrast, to get the atmospheric carbon dioxide level back to safer and more normal levels (300–350 ppm) as quickly as possible, manmade carbon dioxide emissions need to be virtually eliminated, rather than merely stabilized.[p]

That means shifting with all deliberate speed largely to noncarbon and carbon-neutral renewable energy systems while actively removing carbon from the atmosphere into long-term storage and preventing its escape back into the air. Unfortunately, we are nowhere near either a political consensus that this should be done, nor the adoption of an action plan to accomplish it.

We must not assume that just because energy technology will advance, the reduction of carbon emissions will be easier for our children thirty years from now than today. Unless we alter our energy technology mix now to rely more heavily on renewable technologies and to greatly increase the efficiency with which we use energy, our children will have to make drastic cuts in fossil energy use on a shorter timetable and without the benefit of today's still relatively inexpensive fossil fuels for building the alternative global energy system needed. (Fossil fuels are likely to get more expensive as time goes on, as the most easily developed fossil fuel deposits tend to be developed first, and so will have already been exploited.)

Our children's choices will be further constrained by greater population density, greater total energy demand, fewer natural resources, and decades more ecological damage from fossil fuel pollution and climate change. This scenario will put our children at substantially increased risk of runaway global heating. After another generation of business-as-usual, an enlarged and ultimately obsolete fossil fuel energy infrastructure will have to

---

[p] Before human activity greatly increased the carbon dioxide in the atmosphere, natural removal processes were in approximate equilibrium with natural sources of carbon dioxide. Now we are discharging more carbon than the natural removal processes can handle. It should be apparent that if we continue releasing more carbon into an atmosphere already overloaded with it, the time to return to normal carbon levels and to a normal climate will be inevitably prolonged.

be decommissioned and replaced with carbon-free and zero net carbon energy sources, but probably at greater cost than today.

New power plants typically have a 20–40 year operating life. Some survive for 50 years or more. Permitting the construction of new coal- and natural gas-powered generation plants today is thus economically unsound, environmentally irresponsible, and unethical: it commits the world to polluting technologies for decades to come and will confront future generations with the costly dilemma of shutting down working power plants worth billions of dollars before the end of their useful lives—or facing continuing and unacceptable high-carbon emissions and life-threatening climate disturbance.

It is foolish to squander today's resources by locking ourselves into unsustainable fossil fuel technologies instead of investing in climate-friendly technologies that can serve for the foreseeable future. (A detailed description of how to provide needed energy from clean renewable sources is presented in my forthcoming book, *Solving the Climate Crisis: Turning Climate Peril Into Jobs, Prosperity, and Profits.*)

* * *

Myth Eight

Global warming is a blessing in disguise. In fact, it is good for us.

Response:

This is the ultimate contrarian argument. It goes like this. Plants will benefit from increased carbon dioxide levels—after all, carbon dioxide is a nutrient—and with fewer frosts and a longer growing season, agriculture as a whole will benefit. Fish will grow faster and bigger in a warmer world. And people will benefit, as there will be fewer cold winters and fewer deaths from hypothermia. Transportation and communication costs will fall.

These are very simplistic arguments that do not stand up to scrutiny. Climate change *will* have some positive impacts for some people. In northern latitudes, growing seasons will lengthen, some new areas will be suitable for cropping, home heating bills may fall, and people may have to spend less for snow removal. The ice-free shipping season will lengthen in the Arctic and on waterways such as the Great Lakes and the Saint Lawrence Seaway.

On balance, however, these modest benefits do not begin to compensate for the far more devastating socioeconomic and environmental impacts that climate change will bring. For example, soil instability from melting permafrost in the north will counterbalance some of the northern-latitude benefits. And as home heating bills fall in some parts of the land, air conditioning bills will rise in other areas.

As for agriculture, hoped-for increases in crop yields in northern latitudes will be tempered by the unknown impacts of warmer temperatures on precipitation, soil moisture, pests, pathogens, and weeds, and by agricultural losses elsewhere. Furthermore, some of the northern-latitude soils that farmers would like to put into cultivation will turn out to be too light and thin to be of much agricultural value. In addition, as soil temperatures increase, some soils that once were net carbon sinks (where carbon was safely stored) will become sources of additional atmospheric carbon due to shifts in the metabolism of soil microorganisms. These microorganisms will decompose organic matter more rapidly and then release the breakdown products, carbon dioxide and methane, more quickly into the atmosphere. This will further intensify global climate change. A UC Irvine study recently reported that rising temperatures *multiply* the rate of carbon released from soils.[12] This might well damage soil fertility.

As the climate warms, human health and welfare will suffer from the spread of diseases, more frequent heat waves, more severe air pollution, water shortages, violent conflicts over resources, loss of valuable coastal land, and weather-related catastrophes, including hurricanes, droughts, famine, fires, floods, and tornadoes. Thus, the profoundly negative effects of climate change far outweigh the minor reduction in cold-related health effects.

* * *

Myth Nine

Combating climate change would be so expensive that we should wait for more evidence before taking action. Countermeasures would raise energy prices, raise taxes, lower household incomes, reduce business investment, devastate key industries and cost the economy jobs.

Response:

These claims are groundless or grossly exaggerated scare tactics that rest on the "authority" of oil-industry funded studies. Non-industry studies have come to different and often opposite conclusions.[13-19]

One U.S. Department of Energy (DOE) study projected in 2000 that carbon emissions could have been reduced to 1990 levels by 2010 at roughly *zero net cost*.[20] The DOE study did not even include the environmental and public health benefits to the economy from trimming U.S. energy waste and phasing in some low-carbon electricity technologies.

Far from requiring prohibitively expensive carbon taxes, the researchers found that the U.S. could return to its 1990 emissions levels merely by implementing a system of tradable emissions permits valued at $50 per ton of carbon emitted. That would only add 49 cents to the cost of a gallon of gas.[q]

---

[q] U.S. Department of Energy, Energy Information Administration, "Voluntary Reporting of Greenhouse Gases Program - Electricity Factors" (Washington, D.C.: U.S. Energy Information Administration, 2011). www.eia.gov/oiaf/1605/coefficients.html. Accessed May 14, 2012. See also U.S. Department of Energy,

Another research effort conducted by the U.S. DOE and the Environmental Protection Agency (EPA), aided by the Departments of Commerce, Treasury, Labor, and State, found that losses in economic output stemming from even a $100 per ton carbon emission permit system would be small and temporary. No evidence was found that the emissions reduction policy would cause a flight of capital from the U.S.

A third study led by the nonprofit Tellus Institute in 1998 also examined the costs of reducing carbon emissions.[21] The Institute projected that energy use by 2010 could straightforwardly be reduced by 15 percent, and that carbon emissions would then fall by almost 10 percent below 1990 levels.

Instead of imposing unacceptable costs on the economy, they concluded that their scenario would lead to the creation of 800,000 new jobs; the average household would realize energy savings of $530 per year; the gross national product would increase slightly relative to business as usual; and wages and salaries would rise by $14 billion.[22] Tellus forecast that by 2030, greater energy efficiency could reduce U.S. energy consumption by more than 40 percent without any loss of services or comfort.

Where then do the claims of damage to the economy come from? One widely quoted study sponsored by the Global Climate Coalition and funded by the American Petroleum Institute contains exaggerated estimates for the economic damage that would be caused by U.S. adherence to the Kyoto Protocol, including predictions of sharp increases in gasoline prices—a charge that always touches a nerve for the motoring public.[r]

Office of Energy Efficiency & Renewable Energy, 2011, www.fueleconomy.gov/feg/co2.shtml. Accessed May 14, 2012.

[r] Global Climate Coalition. "Climate Economics," March 1999, www.globalclimate.org/economic.htm.

EMISSIONS TRADING

Carbon emissions trading is a market-based process for reducing the release of pollutants into the environment. Here is how it works:

At the start of a trading period, those who must participate are assigned a carbon-dioxide emissions quota or permit (permits may also be auctioned.) Each participant may then either use the allowance by releasing the authorized emissions, or sell the allowance. The issuance of emission permits thus creates a market in emissions and allows participants to trade to their advantage.

An entity with high emissions that lacks a necessary allowance will either have to purchase an unused emissions permit from another party or adopt emission-avoidance technology, whichever is cheaper.

The Protocol authorizes its parties to trade portions of their carbon emission allowances. In the U.S., domestic emissions trading has worked outstandingly in reducing industrial sulfur dioxide emissions—and at far lower cost than industrial polluters had projected.

Based on the unrealistic claim that a huge fee of $241 per ton of carbon would have to be imposed to meet the Kyoto targets, this analysis reached the mistaken conclusion that implementing the Kyoto Protocol would force energy and electricity prices to double, costing the U.S. economy 2.4 million jobs and $300 billion a year. Each household allegedly would have suffered a $2,728 loss in gross domestic product by 2010.

Studies like this tend to be biased, flawed in their design, and thus extremely misleading. Their goal, of course, is not the pursuit of truth or objective social science but the creation of sensational news headlines that create fear of economic hardship, and thus manipulate public opinion against policies designed to reduce climate change. Unfortunately, this tactic often is very effective with uninformed or uncritical members of the public.

Typically, such studies ignore virtually all social and environmental costs of *not* reducing emissions.[s] These studies also often ignore or improperly account for all the indirect benefits of reducing the emissions, such as lower health costs, trade and export benefits from the sale of clean technology, domestic job creation, and gains in worker productivity.

* * *

Myth Ten

Protecting the climate puts an unfair burden on the developed nations to reduce their greenhouse gas emissions without obligating developing nations to do their part. We (the developed nations) should not commit ourselves to act until the developing nations do.

Response:

Total annual carbon emissions from the highly industrialized nations are still far greater than total annual emissions from the developing ones on a per person basis. The U.S. alone produces about 18 percent of the world's carbon emissions with only 4.5 percent of its population.[23] The U.S. and the European Union have produced more than half of all the carbon dioxide emissions produced by humans that have accumulated in the

[s] What, for example, is the cost of disrupting natural life support processes, such as pollination, or the cost of driving more species to extinction, or the cost of more droughts, or wildfires, or the cost of evacuating refugees from flooded areas and resettling them, or of building expensive seawalls and levees, or rebuilding submerged coastal infrastructure?

atmosphere over the past 200 years. China has produced only eight percent.[t]

Emissions from the developing world are surging rapidly, however, and China in 2008 for the first time surpassed the U.S. in carbon dioxide emissions. China now burns more coal than the U.S., India, and Europe combined—three billion tons[24]—and its power sector alone emitted more than three billion tons of carbon dioxide in 2007, largely from the combustion of coal. (An average of two coal-fired power plants are going into operation every week in China, according to the Worldwatch Institute.) The world's fourth-largest carbon emitter is India, which, like China, is increasing its emissions rapidly. India accounts for slightly less than five percent of carbon dioxide emissions, but they have tripled since 1981.

Yet as noted, the developed nations' per person carbon emissions are still far greater than those of developing nations basis, since the population of the developing nations is four times the size of the developed world, and their economies are generally much smaller. For example, India's industrial emissions are roughly 700 pounds per person, and 120 other nations each produce less than 2,200 pounds per person. By contrast, the U.S. annually releases more than 17 tons of carbon per person.

It is therefore neither unfair nor inappropriate for the developed nations to start reducing their carbon emissions. The wealthier nations are not only more responsible for the buildup of atmospheric carbon, but they have also benefited more over time from creating it, and are thus far better able to afford investments to solve the problem. Responsibility for reducing emissions going forward, however, clearly must be shared by both developed and developing nations.

---

[t] See Commonwealth Scientific and Industrial Research Organization Media Release, May 22, 2007, "CO_2 emissions increasing faster than expected," www.csiro.au/news/GlobalCarbonProject-PNAS.html.

Because China and India are growing rapidly, their living standards and aspirations are rising, their energy use is increasing, and their industrial carbon emissions are mounting. If they and the other developing nations do not commit to helping to reduce global emissions, their failure to do so would undermine the world's ability to reduce emissions. The real question is not whether they will participate. They will have to accept emissions limitations eventually. The question is under what terms and what they will exact from the developing nations in return for their cooperation.

If the negotiations are done in an adversarial atmosphere (as seems likely), then the question, "To what degree and at what rate will the developing nations participate with the developed countries in cutting emissions?" will depend on two factors. One is how acutely the developing nations are concerned about the effects that global climate change will have on their nations; the other is how much aid from the developed countries they will require in the form of investment capital (grants or loans) plus the transfer of new energy generating and energy saving technologies.

The developing nations need not knowingly repeat or exceed the mistakes that the developed nations made on their path to industrialization—errors that involved squandering energy and inflicting severe damage on the environment. The newly developing countries could instead leapfrog toward prosperity by both creating and importing cleaner, more energy-efficient technologies and processes than the U.S., Europe and Japan employed. But until the developed nations begin fully meeting their own responsibilities to protect the climate, most developing nations will feel justified in downplaying the problem and avoiding firm commitments.

* * *

Myth Eleven

Adherence to international climate agreements amounts to forfeiting national sovereignty to unelected global authorities. Treaties leave the developed nations hostage to the whims of the developing world.

Response:

These arguments over national sovereignty are supported by those who are willing to partake in the economic benefits of globalization, but not in its responsibilities. The entire world shares the same atmosphere. Chinese, Russian, American, Indonesian, and French carbon dioxide are all commingled as the atmosphere circulates, and so keeping the atmosphere free of harmful gases requires international cooperation.[25]

By engaging in negotiations, we do not forfeit national sovereignty. To the contrary, acting as a sovereign nation, we make beneficial agreements with others that protect our common interests in a healthy climate. To the extent that we choose to accept the limits on the discharge of heat-trapping gases, we do so to accomplish a greater good.

Consider the Kyoto Protocol. Decisions under the protocol, which was signed voluntarily, were made by consensus. No one was forced to comply. The one-nation, one-vote decision process does not mean that large, powerful nations like the U.S. do not have far more informal diplomatic clout than small, less industrialized nations.

If the U.S. tenaciously advocates a position, it is likely to sway other nations and get its way. If its position is not adopted, the U.S. has the option of declining to agree and of thereby blocking consensus on that point. By contrast, if a tiny developing nation objects to a proposal, it is far less likely to convince other signatories to adopt its position, and it may have little recourse but to agree or withdraw.

The charge that the Kyoto agreement would have placed the U.S. and other nations under the governance of unelected global authorities is incorrect. In a democracy, governments routinely delegate authority to negotiators to draft international agreements. Moreover, the U.S. vice president played a significant role in the Kyoto negotiations, and the president strongly endorsed it. Support for the agreement was also strong in other democracies, and in the U.S., a 1999 poll found that 84 percent of the public believed it was important for the U.S. to take action to reduce emissions.

Further Reading

The *New Scientist*'s May 16, 2007 issue had a comprehensive series of articles that appeared under the title, "Climate Change: A Guide for the Perplexed." The articles treat most of the popular climate misconceptions in depth, such as the claim that climate data and computer models can't be trusted and that fluctuations in solar radiation rather than human activity account for climate change. The *New Scientist* also addressed the argument that climate can't be predicted because it's a chaotic system and that warnings about global warming are part of a conspiracy. (See www.newscientist.com/article/dn11462.)

For a technical rebuttal of twenty common myths about global warming, see Brian Angliss's "Anti-Global Heating Claims—A Reasonably Thorough Debunking," at www.scholarsandrogues.com, July 23, 2007. Angliss's article covers many of the same myths as the *New Scientist*, including the mistaken claims that atmospheric concentrations of carbon dioxide are not correlated with atmospheric temperature; that cosmic rays or volcanoes are the true causes of global heating rather than carbon dioxide; that the oceans are cooling rather than warming; and that carbon dioxide concentrations are rising far more slowly than the IPCC claims.

Angliss also provides scientific arguments and technical references refuting the claims that carbon dioxide is too weak a greenhouse gas to cause global heating; that humans aren't the source of increased carbon dioxide concentrations observed in the atmosphere; and that global heating is good for the world. The political, diplomatic and policy results of the oil and coal industry's climate science denial-and-misinformation campaign are summarized in Sharon A. Begley's "The Truth About Denial" (*Newsweek*, August 12, 2007).

UNDETERRED BY FACTS, THE CAMPAIGN CONTINUES

Whereas a decade ago, the oil and coal industries and their allies could get away with denying global warming and with attacking the notion that humans were changing the climate, that strategy is no longer viable. The oil and coal industries and their supporters have nonetheless continued repeating many of their old claims, while updating their campaign with more subtle allegations.

A recent contention has been that, although climate change may be real, and may not be good for the world after all, the most prudent, cost-effective response is to study the problem and defer action. The energy industry and its allies can then always call for more research and set a new date for action even farther in the future. This delaying tactic is heard less and less, however, as the effects of climate change become ever more obvious and severe.

Until people lose patience with this stalling, however, and the deceptive campaigning that goes with it, industries that benefit from fossil fuels will keep fabricating new climate theories and clamoring for research in place of action, and the Earth's climate will continue its grave and alarming deterioration. People in the U.S. and other technologically advanced nations must mobilize enough political support to tackle emissions reduction over industries' objections. Otherwise, real solutions to climate change will remain impossible.

Notes

Chapter 1

1. Joint Science Academy, *Joint Science Academies Statement: Global Response to Climate Change* (Washington, D.C.: The National Academies Press, 2005), www.interacademies.net/File.aspx?id=4825 and nationalacademies.org/onpi/06072005.pdf.
2. Kate Sheppard, "National Association of Manufacturers Claims Climate Bill Would Crush Economy," *Grist Magazine*, August 13, 2009, grist.org/politics/2009-08-12-national-association-manufacturers-climate-bill-crush-economy/, www.accf.org/media/dynamic/3/media_387.pdf.
3. Sharon A. Begley, "The Truth About Denial," *Newsweek*, August 12, 2007, www.newsweek.com/id/32482/output/print.
4. *Ibid.*
5. Peabody Energy, *Energizing the World: 2010 Annual Report* (St. Louis, MO: Peabody Energy, 2010), www.peabodyenergy.com/mm/files/Investors/Annual-Reports/2010BTUAnnualReport.pdf
6. Ross Gelbspan, *The Heat Is On* (Reading, MA: Addison-Wesley Publishing Co., 1997). See also Environmental Media Services and Environmental Information Center, *The 1996 Media Guide to Climate Change*, www.ems.org.
7. Arthur and Zachary Robinson, "Science Has Spoken: Global Warming is a Myth," *The Wall Street Journal*, December 4, 1997.

Chapter 2

1. Chris Mooney, "Some Like It Hot: As the World Burns," *Mother Jones*, May/June 2005.
2. "Put a Tiger in Your Think Tank, Part II: A Survey of ExxonMobil-supported organizations that challenge the scientific basis for concern about global climate change," *Mother Jones*, April 18, 2005 and May/June 2005.
3. Martin Feldstein, "Cap-and-Trade: All Cost, No Benefit" (Washington, D.C: American Enterprise Institute, 2009), www.aei.org/article/energy-and-the-environment/cap-and-trade-all-cost-no-benefit/; Newt Gingrich,"Cap-and-Trade Is

Another Way of Saying 2+2=5" (Washington, D.C: American Enterprise Institute, 2009), www.aei.org/article/energy-and-the-environment/cap-and-trade-is-another-way-of-saying-225/; Kenneth P. Green, "Voters Beware Carbon Tax Seductions" (Washington, D.C.: American Enterprise Institute, 2011). www.aei.org/article/energy-and-the-environment/voters-beware-carbon-tax-seductions/; Kenneth P. Green, "Cap-and-Trade by Any Other Name . . . ." (Washington, D.C.: American Enterprise Institute, 2011), www.aei.org/article/energy-and-the-environment/climate-change/cap-and-trade-by-any-other-name-/

4. Kenneth P. Green, "Dissecting the Carbon Tax" (Washington, D.C.: American Enterprise Institute, 2011), www.aei.org/article/energy-and-the-environment/dissecting-the-carbon-tax/.

5. Samuel Thernstrom, "Beyond Kyoto," *The American*, September 27, 2007.

6. David Callahan, National Committee for Responsive Philanthropy, "$1 Billion for Ideas: Conservative Think Tanks in the 1990s" (Portola Valley, CA.: The Commonweal Institute, 1999).

7. "Patrick J. Michaels, Cato Institute: Policy Scholars" (Washington, D.C.: The Cato Institute, 2010), www.cato.org/people/patrick-michaels.

8. The Cooler Heads Coalition, which was a subgroup of the National Consumer Coalition, is now a coalition of 23 groups, with some international representation. The coalition's website, www.globalwarming.org, is paid for by the Competitive Enterprise Institute, which has received major funding from ExxonMobil. Members include: Alexis de Tocqueville Institution, Americans for Prosperity, Americans for Tax Reform, American Legislative Exchange Council, American Policy Center, America's Future Foundation, Committee for a Constructive Tomorrow, Competitive Enterprise Institute, Fraser Institute, Freedom Works, Frontiers of Freedom, George C. Marshall Institute, Heartland Institute, Independent Institute, Istituto Bruno Leoni (Italy), JunkScience.com, Lavoisier Group (Australia), Liberty Institute (India), National Center for Policy Analysis, Pacific Research Institute, Seniors Coalition, 60 Plus Association, and Small Business and Entrepreneurship Council.

9. "Put a Tiger in Your Think Tank, Part II," *op. cit.*
10. Václav Klaus, "Blue Planet in Green Shackles," (Washington, D.C.: Competitive Enterprise Institute, 2008), www.globalwarming.org
11. Peter Geddes, FREE Vice President, "Future Generations," reprinted from *The Bozeman Daily Chronicle*, May 14, 2003.
12. *The Bozeman Daily Chronicle*, March 14, 2007.
13. "Earth Day, now celebrated around the world, turns 40," *The Cap Times*. host.madison.com/ct/news/local/environment/article_527b9fec-49d5-11df-893d-001cc4c03286.html.
14. *The Bozeman Daily Chronicle*, November 14, 2001.
15. James Inhofe, "We Don't Need A Climate Tax on the Poor," *Wall Street Journal*, June 3, 2008.
16. George Landrith, "Frontiers of Freedom Sends Letter to President Bush on Climate Extortion," April 17, 2008, www.ff.org.
17. Christopher Adamo, May 29, 2008, at www.ff.org.
18. Unsigned. "Put a Tiger in Your Think Tank, Part II," *op. cit.*
19. Tim Ball,"The Science Isn't Settled – The Limitations of Climate Models," March 21, 2007, www.marshall.org.
20. Climate Issues & Questions, 3rd ed. (Washington, D.C.: The Marshall Institute, 2008).
21. Frequently Asked Questions," The Science and Environmental Policy Project, www.sepp.org, June 2008.
22. "Greenhouse Gas Bill Could Raise Gas Prices to $8/Gallon," The National Center for Policy Analysis, Press release, www.nepa.org, June 2, 2008.
23. "Put a Tiger in Your Think Tank, Part II," *op. cit.*
24. The NCPA's website is www.ncpa.org. Its media arm is www.envirotruth.org.
25. The Heartland Institute, www.heartland.org. Accessed June 5, 2008.
26. Harriet Johnson and Joseph L. Bast, *Environment News*, "Climate Change Conference Invigorates Global Warming Debate," May 2008, www.heartland.org.

27. S. Fred Singer (ed.), "Nature, Not Human Activity Rules the Climate" (Chicago, IL: The Heartland Institute, 2008).

28. "Put a Tiger in Your Think Tank, Part II," *op.cit.*

29. Jim Lakely, "Heartland Institute Responds to Stolen and Fake Documents, Press release (email), February 15, 2012.

30. Michael Franc, "Carbon-Cap Conundrum Losing Legislation," www.heritage.org. Accessed June 4, 2008.

31. The Heritage Foundation (staff list), www.heritage.org/about/staff/f/mike-franc.

32. William W. Beach, David W. Kreutzer, Ben Lieberman, and Nicolas D. Loris, "The Economic Costs of the Lieberman-Warner Climate Change Legislation (CDA08-02)," (Washington, D.C.: The Heritage Foundation, May 12, 2008).

33. "Fact Sheet: Heritage Foundation, Exxon Secrets," www.exxonsecrets.org/html/orgfactsheet.php?id=42.

34. "Fact Sheet: Hoover Institution," *op. cit.*, and "Put a Tiger in Your Think Tank, Part II." *op. cit.*

35. Richard W. Behan, "Degenerate Democracy: The Neoliberal and Corporate Capture of America's Agenda," *Public Land Resources Law Review*, vol. 24, 2004, pp. 9-24.

36. S. Fred Singer, "Climate Policy—From Rio to Kyoto: A Political Issue for 2000," EPP 102 DP5 HPEP02FM01 24-05-00 rev1 (Stanford, CA: Hoover Institution on War, Revolution and Peace, 2000).

37. "Put a Tiger in Your Think Tank, Part II," *op. cit.*

38. Peter J. Jacques, Riley E. Dunlap, and Mark Freeman, "The Organization of Denial: Conservative think tanks and environmental skepticism," *Environmental Politics*, vol. 17, no. 3, 2008, pp. 349–385.

39. Office of Atmospheric Programs, U.S. Environmental Protection Agency, *EPA Analysis of the Lieberman-Warner Climate Security Act of 2008* (Washington D.C.: U.S. Environmental Protection Agency, 2008), www.epa.gov/climatechange/downloads/s2191_EPA_Analysis.pdf.

CHAPTER 3

1. "Joseph Fourier, Biography." *Wikipedia*, en.wikipedia.org/wiki/Joseph_Fourier.
2. Spencer Weart and The American Institute of Physics, "The Discovery of Global Warming, The Carbon Dioxide Greenhouse Effect," February 2011, www.aip.org/history/climate/co2.htm#N_3. Accessed April 20, 2012.
3. EPA Analysis of the Lieberman-Warner Climate Security Act of 2008, *op. cit.*
4. *Ibid.*
5. "How do we know that recent CO_2 increases are due to human activities?" See under: Climate science, Paleoclimate, *et al.*, December 22, 2004, at www.realclimate.org.
6. *Ibid.*
7. *Ibid.*
8. James Hansen, Reto Ruedy, Makiko Sato, and Ken Lo, "Global temperature in 2011, trend, and prospects," January 18, 2012, data.giss.nasa.gov/gistemp/2011/.
9. "Urban Heat Island" definition in Susan Mayhew, *Dictionary of Geography* (Oxford, UK: Oxford University Press, 1997), www.oxfordreference.com/pages/Subjects_and_Titles__2E_PS04) and www.Wikipedia.org.
10. British Antarctic Survey, "Press Release - Oldest Antarctic Ice Core Reveals Climate History," British Antarctic Survey, June 9, 2004, www.antarctica.ac.uk/press/press_releases/press_release.php?id=40. Accessed April 16, 2012.
11. IPCC. *Climate Change: 2007*, vol. 1, p. 25.
12. "Global warming threat seen in fertile soil of northeastern US forests, University of California Communications Office, June 11, 2012, today.uci.edu/news/2012/06/nr_soil_120611.php.
13. Robert Pollin, James Heintz, and Heidi Garrett-Peltier, "The Economic Benefits of Investing in Clean Energy (Washington, D.C.: Center for American Progress, and University of Massachusetts, Amherst, Political Economic Research Institute, June 2009).

14. Charles F. Kutscher, American Solar Energy Society, *Tackling Climate Change in the U.S.: Potential Carbon Emissions Reductions from Energy Efficiency and Renewable Energy by 2030*, January 2007.

15. Worldwatch Institute and Center for American Progress, *American Energy: The Renewable Path to Energy Security* (Washington, D.C., 2006).

16. Al Gore, *Our Choice: A Plan to Solve the Climate Crisis* (Emmaus, PA: Rodale Press, 2009.)

17. Arjun Makhijani, *Carbon-Free and Nuclear-Free: A Roadmap for U.S. Energy Policy* (Muskegon, MI: RDR Books; and Takoma Park, MD: IEER Press/ Institute for Energy and Environmental Research Press, 2008).

18. John and Mary Ellen Harte, *Cool the Earth, Save the Economy: Solving the Climate Crisis is EASY*," www.CooltheEarth.us, October 2008.

19. Amory B. Lovins and Rocky Mountain Institute, *Reinventing Fire: Bold Business Solutions for the New Energy Era* (White River Junction, VT: Chelsea Green Publishing, 2011).

20. Interlaboratory Working Group, *Scenarios for a Clean Energy Future* (Oak Ridge, TN; Oak Ridge National Laboratory and Berkeley, CA.: Lawrence Berkeley National Laboratory, 2000), ORNL/CON-476 and LBNL-44029.

21. Tellus Institute, and Stockholm Environment Institute-Boston, *Policies and Measures to Reduce CO_2 Emissions in the United States: An Analysis of Options through 2010* (Boston and Stockholm, 1998).

22. *Ibid.*

23. U.S. Department of Energy, Energy Information Administration, *Country Analysis Brief*, www.eia.gov/countries/country-data.cfm?fips=US&trk=p1#cde.

24. Bill McKibben, "Can China Go Green, *National Geographic*, June 2011.

25. Tony Blair and The Climate Group, *Breaking the Climate Deadlock: A Global Deal for Our Low-Carbon Future* (London, UK: Office of Tony Blair). Report submitted to the G8 Hokkaido Toyako Summit, June 2008.

Appendix

Important Climate Websites and other Information Sources

An ocean of high-quality information is available about global climate change and renewable energy, on the web and in print. The problem isn't how to find it, but how to avoid *drowning* in it and how to know what information is reliable and what is incorrect or intended to mislead.

The most comprehensive scientific reports about climate change are those of the United Nations Intergovernmental Panel on Climate Change, which are available on the web at www.ipcc.ch/ and in print from Cambridge University Press.

Useful links from the IPCC site include:

- Global Environment Facility (GEF)
- IPCC Data Distribution Centre
- United Nations (UN)
- United Nations Environment Programme (UNEP) (Geneva, Nairobi)
- United Nations Framework Convention on Climate Change (UNFCCC)
- UN Gateway to Climate Change
- World Meteorological Organization (WMO)

By going, for example, to Linkages on the International Institute for Sustainable Development's website, www.iisd.ca/linkages-update, you will find links to the most important international negotiations on global climate change that have occurred since 1997, as well as to authoritative research on a vast array of important climate and energy issues.

The Secretariat of the UN Framework Convention on Climate Change, www.unfccc.de—the umbrella agreement that led to the Kyoto Protocol on greenhouse (GHG) gas reduction—lets users keep up with international climate negotiations and also offers a searchable database of country-by-country GHG information. You may also want to visit the sites of the UN Environment Programme (UNEP), www.unep.ch/iucc. A portion UNEP's site has the official documents of the intergovernmental climate negotiations, sessions of which are known as "Conference(s) of the Parties." A NASA master directory of worldwide climate change data holdings (the *Global Change Master Directory*) can be found at www.gcmd.gsfc.nasa.gov.

The United States Environmental Protection Agency maintains an enormously useful web site at www.epa.gov/climatechange/index.html with lots of information on greenhouse gas emissions and climate science. You can find global, national, and individual facility emission data as well as information on how you or your business can reduce your GHG emissions. The site also links to detailed information on climate change impacts, adaptation to climate change, and EPA's policy efforts and scientific research as well as to relevant laws and regulations. Once you are connected to a major climate change web site like this, by following its electronic links to other web destinations, you can generally learn anything known to science about climate and energy.

The United States Department of Energy's (DOE's) Energy Information Administration at www.eia.gov has data on energy-related GHG emissions and is a great source of information on the major energy sources and energy technologies, especially current trends and historical data on production of electricity and fuels as well as renewable energy sources. You'll find energy units and concepts explained here as well as international energy data. If you haven't visited it, the range and depth of the information there will amaze you.

For information focused specifically on renewable energy and energy efficiency, the DOE's National Renewable Energy Laboratory (www.nrel.gov) is the place to go both for scientific and technical information as are the web sites of other national laboratories, such the Lawrence Berkeley Laboratory, Sandia National Laboratory, and Oak Ridge National Laboratory, host of the Carbon Dioxide Information Analysis Center. (The DOE currently has a total of 17 national laboratories.) DOE's Energy Efficiency and Renewable Energy Clearinghouse at www.eere.energy.gov is also valuable.

The National Energy Information Center (NEIC), a branch of the EIA, provides inquiry services and distributes all EIA information products, including data publications, interpretive reports, directories, and EIA press releases. The Center's inquiry unit answers questions over the telephone or refers callers to the appropriate source for information. NEIC also offers free energy information on a variety of topics, including educational materials for teachers. Its phone number is (202) 586-8800, the website is www.eia.gov/neic/neicservices.htm, and their email is InfoCtr@eia.doe.gov.

You can contact NEIC for—

- Customized responses to information and data requests.
- Expert navigation of the EIA web site to reach relevant EIA data and reports.
- Assistance with explanation and interpretation of data and conversions.
- Referrals to EIA experts for in-depth data and information requirements.
- Print versions and subscriptions to EIA reports.
- Copies of selected historical data that are unavailable in electronic format.

- Referrals to DOE and other Federal agencies, state agencies and trade associations for relevant information.

In addition to the government sites already noted, other important Federally funded resources include the National Center for Atmospheric Research, www.ucar.edu, the National Climatic Data Center, www.ncdc.noaa.gov, the National Oceanic and Atmospheric and Administration, www.noaa.gov, and the National Aeronautics and Space Administration Goddard Institute for Space Studies, www.giss.nasa.gov/. The Goddard Institute has extensive data sets for professional climatological research and other useful resources, including feature articles, scientific briefs, and research reports.

The U.S. Global Climate Change Research Program, set up by Presidential initiative, has a wealth of information at www.globalchange.gov that integrates federal research on global change and climate change. This is a *must visit* site that not only contains results of ongoing national assessments of the potential impacts of climate change—including many regional and sectoral studies—but alo offers links to most other federal climate-related programs. These include the federal government's Climate Change Adaptation Task Force, National Ocean Council, Climate Change Technology Program, Subcommittee on Disaster Reduction, and Interagency Task Force on Carbon Capture and Storage.

As an example, here are some of the offerings you can find under just one of these headings, the U.S. Climate Change Technology Program:

- *Inventory of Greenhouse Gas Reducing Technology Deployment Activities*

- *Strategies to Promote Commercialization and Deployment of GHG Intensity-Reducing Technologies and Practices*

- *Assessing Economic Impacts of Greenhouse Gas Mitigation – Summary of a Workshop*

Most state energy commissions and public utilities commissions are also on-line; see in particular the California Energy Commission's site at www.energy.ca.gov. The Department of Energy has information about renewable and energy efficiency as well as federal, state, local, and tribal grant programs at www.eere.energy.gov/topics/government.html and the Interstate Renewable Energy Council at www.irecusa.org has a database of information on more than 2,700 state, local, utility, and federal incentives and policies that promote renewable energy and energy efficiency. Sustainable Minnesota at www.me3.org also has state renewable energy information

To keep up with new developments about climate change, subscribe to the International Institute for Sustainable Development's comprehensive biweekly electronic digest of climate news from the world press, "Climate News," at www.iisd.ca/process/climate_atm.htm. IISD also provides the *Earth Negotiations Bulletin* covering international climate change negotiations at http://www.iisd.ca/.

Leading national environmental and other nongovernmental organizations active on energy and climate issues, such as Greenpeace, the Natural Resources Defense Council, and the Sierra Club, are listed below. They can be found on the web and in the comprehensive *Conservation Directory*, published by the National Wildlife Federation of Washington, D.C., available in print as well as electronically.

AUTHORITATIVE REBUTTALS TO COMMON CLIMATE MYTHS

Climate Central – www.climatecentral.org

Real Climate – www.realclimate.org

Skeptical Science – www.skepticalscience.com/argument.php

SCHOLARLY CRITIQUES OF CLIMATE CHANGE DENIAL

Riley E. Dunlap and Aaron McCright, Chapter 14, "Climate Change Denial: Sources, Actors, and Strategies" in Constance Lever-Tracy (ed.), *Routledge Handbook of Climate Change and Society* (Oxon, UK and New York, NY: Routledge, 2010).

Riley E. Dunlap and Aaron McCright, Chapter 10, "Organized Climate Change Denial," in John S. Dryzek, Richard B. Norgaard, and David Schlosberg (eds.), *The Oxford Handbook of Climate Change and Society* (Oxford and New York: Oxford University Press, 2011).

William D. Nordhaus, "Why the Global Warming Skeptics Are Wrong," *The New York Review of Books*, March 22, 2012.

Karie Marie Norgaard, Chapter 27, "Climate Denial: Emotion, Psychology, Culture, and Political Economy" in John S. Dryzek, Richard B. Norgaard, and David Schlosberg (eds.), *The Oxford Handbook of Climate Change and Society* (Oxford and New York: Oxford University Press, 2011).

Orin H. Pilkey, and Keith C. Pilkey, Chapter 3, "Doubts, Uncertainties, and Qualms," and Chapter 4, "The Manufacture of Dissent," in *Global Climate Change: A Primer* (Durham and London: Duke University Press, 2011).

Naomi Oreskes, "The Scientific Consensus on Climate Change: How Do We Know We're Not Wrong?" in Joseph F.C. DiMento and Pamela M. Doughman, *Climate Change: What It Means for Us, Our Children, and Our Grandchildren* (Cambridge, MA.: The MIT Press, 2007), pp. 65–66.

NOTEWORTHY BOOKS ON CLIMATE CHANGE DENIAL

John Cook and Haydn Washington, *Climate Change Denial: Heads in the Sand,* (Oxon, UK and New York, NY: Routledge, 2011).

James Lawrence Powell, *The Inquisition of Science* (New York: Columbia University Press, 2011).

CLIMATE CONTRARIANS

The American Petroleum Institute
www.api.org

Global Climate Coalition
www.globalclimate.org

The Heartland Institute
www.ClimateWiki.org

The World Climate Report
www.greeningearthsociety.org/climate

Western Fuels Association
www.westernfuels.org

NATIONAL ORGANIZATIONS AND MEDIA WORKING ON OR PROVIDING COVERAGE OF ENERGY AND CLIMATE ISSUES

AlertNet
www.alertnet.org/climate

Alliance to Save Energy
www.ase.org

American Council for an Energy Efficient Economy
www.aceee.org

American Council on Renewable Energy
www.acore.org

Alliance for Climate Protection
www.climateprotect.org

Alliance for Renewable Energy
www.allianceforrenewableenergy.org

American Forests
www.americanforests.org

American Solar Energy Society
www.ases.org

Business Council for Sustainable Energy
www.bcse.org

KQED News Climate Watch
blogs.kqed.org/climatewatch

Clean Power Campaign
www.cleanpower.org

Center for Energy Efficiency and Renewable Technology
www.ceert.org

Center for Environmental Information, Inc. (Rochester, NY)
www.rochesterenvironment.com/resources.htm

Center for Resource Solutions
www.resource-solutions.org

Climate Science Legal Defense Fund
climatesciencedefensefund.org/press

Climate Ark
www.climateark.org

Climate Progress
thinkprogress.org/climate/issue

Cllimate Reality Project
www.climaterealityproject.org

Climate Science Rapid Response Team
www.climaterapidresponse.org

Climate Central
www.climatecentral.org
(Author's Note, this is a phenomenal website for obtaining scientific information on climate change.)

Climate Action Network
www.climatenetwork.org

Corporate Watch
www.corpwatch.org

Critical Mass Energy Project
www.citizen.org/cmep

Earth Day 2000
www.earthday.net

Earth Island Institute
www.earthisland.org

Environmental Alliance for Senior Involvement
www.easi.org

Environmental Defense Fund
www.edf.org

Environmental and Energy Study Institute
www.eesi.org

Environmental Media Services
www.ems.org

Friends of the Earth
www.foe.org

Greenpeace
www.greenpeace.org

Global Climate Change Digest **(published in** ***Global Change,*** **the electronic edition)**
www.globalchange.org/default.htm

Institute for Energy and Environmental Research
www.ieer.org

International Climate Change Partnership
www.iccp.net

Midwest Renewable Energy Association
www.midwestrenew.org/home

National Audubon Society
www.audubon.org

National Wildlife Federation
www.nwf.org

Natural Resources Defense Council
www.nrdc.org

Tiempo Climate Newswatch
www.tiempocyberclimate.org/newswatch

OSS Foundation
www.ossfoundation.us

Pace Law Energy and Climate Center
www.law.pace.edu/energy-and-climate-center

Pacific Institute for Studies in Development, Environment, and Security
www.pacinst.org

Center for Change and Energy Solutions
www.c2es.org

Physicians for Social Responsibility
www.psr.org

The Federation of State Public Interest Research Groups
www.uspirg.org

Real Climate
www.realclimate.org

Renewable Energy Policy Project
www.repp.org

The Rocky Mountain Institute
www.rmi.org

The Rural Alliance for Renewable Energy
www.infinitepower.org/rare

Sierra Club
www.sierraclub.org

Solar Century
www.solarcentury.co.uk

Think Progress
www.thinkprogress.org/climate/issue

University of East Anglia (England) Climate Research Unit
www.cru.uea.ac.uk

Union of Concerned Scientists
www.ucsusa.org

U.S. Climate Action Network (same as CAN)
www.usclimatenetwork.org

Western Clean Energy Campaign
www.westerncec.org

World Wildlife Federation
www.wwf.panda.org/about_our_earth/aboutcc

Worldwatch Institute
www.worldwatch.org

SELECTED GOVERNMENTAL ORGANIZATIONS

International Energy Agency (IEA)
www.iea.org

Natural Resources Canada
climatechange.nrcan.gc.ca

The State of California Air Resources Board
www.arb.ca.gov/cc/cc.htm

The State of California Climate Change Portal
www.climatechange.ca.gov

The National Oceanic and Atmospheric Administration ClimateWatch magazine
www.climate.gov/#climateWatch

United States Global Change Research Program
www.globalchange.gov

The University Corporation for Atmospheric Research
www2.ucar.edu

USDOE, Energy Efficiency and Renewable Energy
www.eere.energy.gov

USDOE, National Center for Photovoltaics
www.nrel.gov/ncpv

USDOE, National Renewable Energy Laboratory
www.nrel.gov

USDOE, Sandia National Laboratories – Energy, Climate, & Infrastructure Security
www.energy/sandia.gov

UN-Secretary-General, Sustainable Energy for All
www.sustainableenergyforall.org

Renewable Energy Industry Organizations

American Hydrogen Association
www.americanhydrogenassociation.org

American Wind Energy Association
www.awea.org

Export Council for Energy Efficiency
www.ecee.org

Geothermal Education Office
www.geothermal.marin.org

Geothermal Energy Association
www.geotherm.org

Geothermal Resources Council
www.geothermal.org

National Biodiesel Board
www.biodiesel.org

National Hydropower Association
www.hydro.org

International Geothermal Association
www.geothermal-energy.org

Renewable Energy Alliance
www.renewableenergy21.com

Renewable Energy Business Network
rebn-east.weebly.com

Renewable Fuels Association
www.ethanolrfa.org

Rural Renewable Energy Alliance
www.rreal.org

Solar Energy Industries Association
www.seia.org/main.htm

Southern Alliance for Clean Energy
www.cleanenergy.org

Womens' Council on Energy and the Environment
www.wcee.org

Regional Energy and Climate Information

Environmental Law and Policy Center (Midwest)
www.elpc.org

Northwest SEED Coalition (Sustainable Energy for Economic Development)
www.nwseed.org

SEED Coalition (Texas)
www.seedcoalition.org

SEED Coalition (Massachusetts)
www.masstech.org/seed

Tennessee *Renewable Energy* and *Economic Development* Council
www.treedc.us

University of North Carolina Institute for the Environment, Center for Sustainable Energy, Environment and Economic Development
www.ie.unc.edu/cseeed/index.cfm

ELECTRIC UTILITY INDUSTRY ORGANIZATIONS

Edison Electric Institute
www.eei.org

Electric Power Research Institute
www.epri.com

North American Electric Reliability Council
www.nerc.com

SOLAR ENERGY WEB SITES

American Solar Energy Society
www.ases.org

National Renewable Energy Laboratory
www.nrel.gov

Solar Energy Industries Association
www.seia.org

Solar Energy International
www.solarenergy.org

INTERNATIONAL ORGANIZATIONS AND PROGRAMS WITH ENERGY AND CLIMATE CONCERNS

Earth Negotiations Bulletin
www.iisd.ca/enbvol/enb-background.htm

Global Energy Network Institute
www.geni.org

Global Environment Facility (a World Bank program)
www.gefweb.org

Intergovernmental Panel on Climate Change
www.ipcc.ch

International Climate Change Partnership
www.iccp.net

International Council for Local Environmental Initiatives
www.iclei.org

International Institute for Sustainable Development
www.iisd.ca

International Energy Agency
www.iea.org

International Solar Energy Society
www.ises.org/index/html

Organization for Economic Cooperation and Development
www.oecd.org

Solar Electric Light Fund
www.self.org

Solar Energy International
www.solarenergy.org

Stockholm Environment Institute
www.sei.se

Tata Energy Research Institute
www.teriin.org

United Nations Environment Programme (UNEP) Collaborating Centre on Energy and the Environment
www.unep.org

U.S. Country Studies Program
www.gcrio.org/CSP

World Climate Research Program (see World Meteorological Organization below)
www.wmo.int/pages/index_en.html

World Energy Council
www.worldenergy.org

World Meteorological Organization
http://www.wmo.int/pages/index_en.html

World Health Organization
www.who.int/en

Scientific Journals That Cover Climate Change and the Environment

Atmospheric Chemistry and Physics

Atmospheric Environment

Climate Dynamics

Climate Policy

Climate Research

Climatic Change

Energy Policy

Environmental Research Letters

Environmental Resource Economics

Environmental Science and Technology

Geophysical Research Letters

Global Change Biology

Journal of Climate

Journal of Geophysical Research

Nature

Nature Geoscience

Nature Reports Climate Change

Proceedings of the National Academy of Sciences

Science

Major Scientific Societies That Cover Climate and Related Environmental Issues

American Geophysical Union

American Institute of Professional Geologists

American Meteorological Society

American Sociological Association

American Solar Energy Society

Association of American Geographers

Association of Climate Change Officers

Association of Environmental & Engineering Geologists

Development Studies Association

Ecological Society of America

Environmental & Engineering Geophysical Society

Environmental and Energy Study Institute

European Association of Geoscientists & Engineers

European Geosciences Union

European Wind Energy Association

Geochemical Society Geological Society (London, UK)

Geological Society of America

International Society for Environmental Biogeochemistry

International Association for Urban Climate International

Association of Meteorology and Atmospheric Sciences

International Society for Ecological Economics

The Royal Society Oceanography Society

The World Conservation Union

ACKNOWLEDGMENTS

I'm very grateful to Wendy Li, Jennifer Millman, Irene Saunders, Maria Terekhov, Ashley Warner, and Ping Wu for research assistance and to Dennis Martz for production assistance.

I appreciate the diligent professional editing provided by Dr. Marta Tanrikulu of TanMar Editorial, and the eye-catching book cover design by artist and designer Chii Maene of Berkeley, CA., who gracefully put up with many design changes. My thanks as well to all the friends and family who provided input on the cover.

Research assistance on portions of a predecessor volume, *Beating the Heat*, later adapted for *Climate Myths*, came from researchers Rachel Anderson, Elsa Lai Fan, Jan Thomas, and Andrew Perkins. Paul Hesse and Eric Wesselman also assisted with research for its Appendix, while Maria Terekhov helped identify needed climate experts. Expert content advice for that volume came from Langston James Goree, VI, Chad Carpenter, Professor John Harte, Dr. Andrew Horner, Peter Kelley, Dr. Jon Koomey, Dr. Douglas N. Koplow, Dr. Florentine Krauss, Dan Lashof, Dr. Mark Levine, Alan Sanstad, Dr. Matthias Schabel, and the late Professor Stephen H. Schneider. Editorial advice came from David Dunaway, Bob Hall, Tom Brenner, Jerry Kramer, Robert Masterson, Alison Monroe, and Max Tomlinson. The author of course takes full responsibility for any errors in the current text.

Climate research support was generously provided by the late Newton D. Becker of the Newton D. and Rochelle F. Becker Foundation. Newt not only supported the research for *Beating the Heat*, but also distributed hundreds of copies for educational purposes. I regret that he cannot be here to receive a copy of *Climate Myths* or the other two forthcoming climate books (*Climate Peril* and *Climate Solutions*) to which the research he supported has contributed.

Index

A

B

About the Author

Dr. John J. Berger is the author and editor of 11 books on climate, energy, and natural resources. He is a graduate of Stanford University and has a master's in energy and natural resources from UC Berkeley and a Ph.D. in ecology from UC Davis. He has been a journalist, professor, and leader of national environmental organizations. He has also served as a consultant on energy and natural resources to government, scientific, academic, and nonprofit organizations, including the U.S. Congress and the National Academy of Sciences.

Contributors' Biographical Information

John H. Adams is Founding Director of the Natural Resources Defense Council (NRDC) and Chair, Open Space Institute. He is also a past President, past Executive Director, and past Chairman of NRCD. He co-founded NRDC in 1970, serving as NRDC's executive director and later as its president from its inception until 2006—a tenure unparalleled by the leader of any other environmental organization. As NRDC's founding director, Adams continues to play an active role in the organization on the local, national, and international levels. In February 2011, Adams received the Presidential Medal of Freedom—our nation's highest civilian honor—from President Obama. In announcing the award, the president referenced Rolling Stone Magazine's description of Adams, saying: "If the planet has a lawyer, it's John Adams."

In 2010, Adams and his wife Patricia co-authored *A Force for Nature*, a memoir recounting their forty years of battles and victories with NRDC. He is currently chair of the board of the Open Space Institute and sits on the boards of numerous other environmental organizations. He has also served on governmental advisory committees, including President Clinton's Council for

Sustainable Development. Prior to his work at NRDC, Adams served as assistant U.S. attorney for the Southern District of New York. He is a graduate of Michigan State University and the Duke University School of Law. Adams lives in the Catskill Mountains of New York, not far from the farm where he was raised.

Dr. John Harte holds a joint professorship in the Energy and Resources Group and the Ecosystem Sciences Division of the College of Natural Resources. He received a BA in physics from Harvard University in 1961 and a PhD in theoretical physics from the University of Wisconsin in 1965. He was an NSF Postdoctoral Fellow at CERN, Geneva, during 1965–66 and a Postdoctoral Fellow at the University of California, Lawrence Berkeley Laboratory, during 1966–68. During the next 5 years, he was Assistant Professor of Physics at Yale University and has been at Berkeley since 1973. Harte is a Fellow of the American Physical Society, and in 1990 was awarded a Pew Scholars Prize in Conservation and the Environment. In 1993 he was awarded a Guggenheim Fellowship and was elected to the California Academy of Sciences. In 1998 he was appointed a Phi Beta Kappa Distinguished Lecturer and a Distinguished Ecologist Lecturer at Colorado State University. He is the 2001 recipient of the Leo Szilard prize from the American Physical Society, the 2004 recipient of the UC Berkeley Graduate Mentorship Award, and in 2006 received a Miller Professorship. He has served on six National Academy of Sciences Committees and has authored over 190 scientific publications, including eight books, on topics such as biodiversity, climate change, biogeochemisty, and energy and water resources. They include: *Maximum Entropy and Ecology: A Theory of Abundance, Distribution, and Energetics* (Oxford University Press, 2011) and *How to Cool the Planet (and Save the Economy, Too).*

Harte's research focuses on the effects of human actions on, and the linkages among, biodiversity, ecosystem structure and function, and climate. His work spans a range of scales, from plot to landscape to global, and utilizes field manipulation experiments, the study of patterns in nature, and mathematical modeling. Two specific goals are to understand the nature and causes of patterns in the distribution and abundance of species and to understand the extent to which ecosystem responses to climate change may result in feedbacks to climate that can either ameliorate or exacerbate global warming. An overarching goal of his research is to understand the interdependence of human well-being and the health of ecosystems.

Dr. Kevin E. Trenberth is a distinguished senior scientist in the Climate Analysis Section at the National Center for Atmospheric Research. From New Zealand, he obtained his Sc. D. in meteorology from the Massachusetts Institute of Technology. He has been prominent in most of the Intergovernmental Panel on Climate Change (IPCC) scientific assessments of climate change and has also extensively served the World Climate Research Programme (WCRP) in numerous ways. He now chairs the WCRP Global Energy and Water Exchanges (GEWEX) project. He has also served on many U.S. national committees. He is a fellow of the American Meteorological Society, the American Association for the Advancement of Science, the American Geophysical Union, and an honorary fellow of the Royal Society of New Zealand. He has published over 480 scientific articles or papers, including 47 books or book chapters, and over 213 refereed journal articles and has given many invited scientific talks as well as appearing in a number of television and radio programs and newspaper articles. He is listed among the top 20 authors with the highest citations in all of geophysics. See www.cgd.ucar.edu/cas/trenbert.html.

CPSIA information can be obtained at www.ICGtesting.com
Printed in the USA
LVOW081003120513

333369LV00001B/76/P